Thermotropic Liquid Crystals

Critical Reports on Applied Chemistry Volume 22

Thermotropic Liquid Crystals

edited by G.W. Gray, FRS

Published on behalf of the Society of Chemical Industry by
John Wiley & Sons
Chichester · New York · Brisbane · Toronto · Singapore

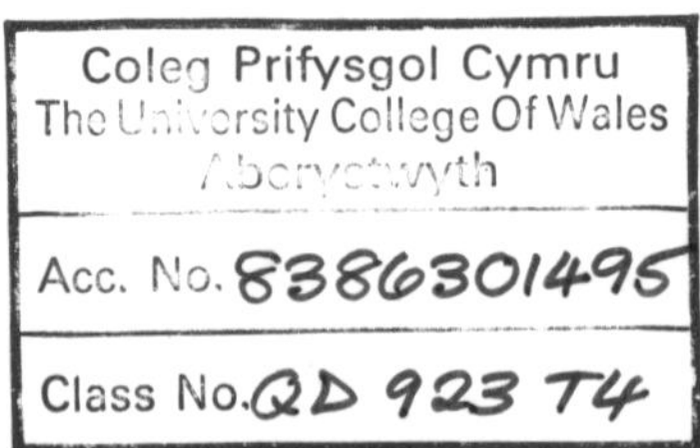

Library of Congress Cataloging-in-Publication Data:
Thermotropic liquid crystals.
(Critical reports on applied chemistry; v. 22)
Includes index.
Contents: Structural classification of liquid crystals/A.J. Leadbetter — Liquid crystal behaviour in relation to molecular structure/ K.J. Toyne — Material requirements for nematic and chiral nematic electro-optical displays/ D. Coates — [etc.]
1. Liquid crystals. I. Gray, G.W. II. Society of Chemical Industry (Great Britain III. Series.
QD923.T47 1987 548′.9 87-2023

ISBN 0 471 91504 1

British Library Cataloguing in Publication Data:
Thermotropic liquid crystals. — (Critical reports on applied chemistry; v. 22).
1. Liquid crystals
I. Gray, G.W. II. Society of Chemical Industry III. Series
548′.9 QD923

ISBN 0 471 91504 1

Printed and bound in Great Britain by
Biddles Ltd, Guildford and King's Lynn

Contents

List of contributors

D. Coates†	*STC Technology Ltd, Standard Telecommunication Laboratories, London Road, Harlow, Essex, CM17 9NA, England*
G. W. Gray	*Department of Chemistry, The University, Hull, HU6 7RX*
H. Finkelmann	*Institut für Makromolekulare Chemie, Hermann-Staudinger-Haus, Universität Freiburg, Stefan-Meier-Strasse 31, D-7800 Freiburg i. Br., West Germany*
A. J. Leadbetter	*Rutherford Appleton Laboratory, Chilton, Didcot, Oxfordshire, OX11 0QX, England*
D. G. McDonnell	*Royal Signals and Radar Establishment, Malvern, Worcestershire, WR14 3PS, England*
I. Sage	*BDH Limited, Broom Road, Poole, Dorset, BH12 4NN, England*
K. J. Toyne	*Department of Chemistry, University of Hull, Hull, HU6 7RX, England*

†Present address BDH Limited, Broom Road, Poole, Dorset BH12, 4NN, England

Editor's introduction

Liquid crystals have experienced an explosive growth in the last twenty years. This has arisen primarily because successful applications of liquid crystals have been developed, particularly in the area of electrooptical displays. The consequent demand for stable liquid crystal materials manifesting carefully controlled combinations of a range of physical properties — viscosity, elastic properties, dielectric properties, response times, etc. — in the required liquid crystal phase, not only at ambient temperatures but also over temperature ranges as wide as -40 to $+100\,°C$, has acted as a spur for materials research by organic chemists working in close collaboration with physicists, both pure and applied, and with electrical and electronic engineers. Not surprisingly, this intensive materials programme has quite frequently brought to light new materials with hitherto unknown liquid crystal phases, and this in its turn has stimulated structural studies of these systems by microscopic, calorimetric, X-ray and neutron-scattering techniques. In turn, some of these novel phases have features of very considerable theoretical significance. Added to all these developments, the last ten years have also witnessed the development and study of liquid crystal polymers, and the potential of these materials in the field of applications is of great interest. Main chain liquid crystal polymers are already well known for their ability to form very high tensile strength fibres, and the side chain liquid crystal polymers are under active investigation as materials for data storage systems.

A direct outcome of all this activity in research and in commercial exploitation has been that the subject of liquid crystals has become known to an increasing audience of scientists from a range of disciplines, and a need has developed among such scientists for texts on the subject which will enlighten them further and assist them if they wish to become involved in research in the area. Specialist texts in the area certainly do exist, but these were never written for enquiring beginners in the field, and the problem is compounded since reviews in the multidisciplinary area of liquid crystals are disseminated over a wide range of journals from different disciplines.

I was therefore extremely interested when I was asked if I would be prepared to act as Senior Reporter and Editor of a book reviewing 'Liquid Crystals' with the object of presenting to chemists the 'state of the art' in the chemistry of liquid crystals. I saw this as an opportunity to produce, at least for chemists, a good introductory text which would provide important background information and be useful to chemists interested in learning more about the subject. I was immediately faced with the problem that the subject is now very large and that within the limits of the series of Critical Reports on Applied Chemistry it would clearly be impossible to cover all aspects.

Liquid crystals can roughly be divided into two areas:

1. *Lyotropic liquid crystals* which are formed from compounds with amphiphilic properties and solvents (commonly water). Between the pure amphiphile and its isotropic solution in an excess of the solvent there exist, at intermediate concentrations, structured phases, consisting of ordered arrangements of micelles composed of amphiphile and solvent, which exhibit anisotropic properties characteristic of liquid crystal phases. Common examples of such lyotropic liquid crystals are those produced from soaps and other detergent systems and water.
2. *Thermotropic liquid crystals* which are formed from compounds (predominantly organic, but also organometallic) whose molecules are mainly either rod-shaped or disc-shaped, either by heating the crystalline solid or by cooling the isotropic liquid, i.e. by thermal effects. Several materials exist in a liquid crystal phase at room temperature, and many other cases of this type can be produced by using mixtures of compounds. All components of the mixture need not necessarily be individually capable of forming liquid crystal phases or mesophases, i.e. need not be mesogens. Non-mesogenic components may often be added to depress the melting point or to tune other physical parameters such as viscosity.

Since most of the interesting applications of liquid crystals have involved those of the thermotropic type, lyotropic systems are omitted. This was a decision dictated by necessity and in no sense implies that the importance of lyotropic liquid crystals and research in this area is underrated. A useful text for chemists could well be devoted just to lyotropic liquid crystals, but in the prevailing circumstances the present text concentrates on the thermotropic area and so became entitled *Thermotropic Liquid Crystals*.

The book consists of six chapters: there now follow brief comments on each chapter, largely to draw attention to the authors and their objectives.

Chapter 1. Structural classification of liquid crystals (A.J. Leadbetter)

One of the greatest problems encountered by a newcomer to the field of liquid crystals (and indeed it is also a problem for some who have worked in the field for some time) is the wide diversity of liquid crystal phases — particularly the many phases of the smectic type — that are known, and even today continue to be discovered. Chapter 1 sets out to define clearly how liquid crystal phases of the thermotropic kind are classified and what for each are its unique structural features. It will be a most welcome source for understanding clearly the subtle differences between the many phases of the truly smectic and crystal types.

Chapter 2. Liquid crystal behaviour in relation to molecular structure (K.J. Toyne)

An overview of the organic chemistry of the materials that form liquid crystal phases of the thermotropic type, particularly those of direct relevance to applications, is provided. The organic chemistry of liquid crystal materials is often referred to as simple by those who know little about the subject. The molecules must be relatively simple to be of interest, otherwise their phases would not be produced until very high temperatures, so making the materials of little value. Moreover, the properties of these systems must be 'tuned' very carefully to meet the requirements for specific applications, and so the relatively simple molecules must be capable of incorporating just the right features to give a desirable viscosity for the correct type of liquid crystal phase, a high or a low dielectric anisotropy, a high or a low birefringence, etc. Purities in excess of 99.7 per cent are also essential. The molecules are therefore simple but beautiful. Chapter 2 gives a nicely balanced overview of the systems involved, ranging today from the purely aromatic or heteroaromatic to the purely alicyclic, and incorporating many interesting ring types and structural features.

Chapter 3. Materials requirements for nematic and chiral nematic electrooptical displays (I. Sage)

This chapter sets out to explain why liquid crystals have become so important in the field of electrooptical displays. The basics and subtleties of the twisted nematic display device, which features in most everyday liquid crystal displays and upon which the success of the liquid crystal display industry has been based, are discussed fully before leading on to more recently developed devices using nematics and chiral nematics, and discussing their potential for the future. Particular emphasis is placed on the requirements imposed by the different display modes on the liquid crystal materials to be used in the device.

Chapter 4. Materials requirements for smectic liquid crystal displays (D. Coates)

For a long time, electrooptical display applications of liquid crystals were concentrated upon nematic and chiral nematic liquid crystals, to the neglect of those of the smectic type. When this text was first planned, only laser addressed smectic A displays were commonly under consideration, but interest was becoming greater in fast switching ferroelectric chiral smectic C displays. Due to the rapidity of developments in the field, in the time span of bringing together this text, an extremely interesting electrically addressed smectic A device has been launched by Standard Telecommunication Laboratories at Harlow, and the above-mentioned ferroelectric displays, with their potential to switch at video frame rates, have become the most active current area of applications research in the entire field. This chapter sets out to discuss these smectic liquid crystal displays and to elaborate their future significance in the field of large area displays with storage capabilities. The author had the vision with his colleagues, to see the virtues in smectic materials at a time when most others were concentrating on nematics and chiral nematics.

Chapter 5. Thermochromic cholesteric liquid crystals (D.G. McDonnell)

For many years it has been known that cholesteric or chiral nematic liquid crystals possess a twisted helical structure which is capable of responding to temperature change through change in the helical pitch length. As a consequence, the wavelength of the light reflected from the planar cholesteric structure changes with temperature, and if the reflected light spans the visible range, the eye will see distinct changes in colour as the temperature varies. This properly opens up many possible applications in the field of surface thermography and thermooptics. Original materials studied in this area were cholesteryl esters of less than desirable stability and durability, and as a consequence, devices developed had a less than desirable permanence, but new materials have radically changed this situation.

Chapter 6. Liquid crystal polymers (H. Finkelmann)

Liquid crystal polymers both of the main chain and side chain types are exciting a great deal of interest. Nematic, smectic, cholesteric and discotic analogues of low molar mass liquid crystals are now known, and both thermotropic and lyotropic systems may be prepared. The existence of these materials widens the sphere of interest of the liquid crystal chemist to that of polymer chemistry with all its unique specialisms and problems. This chapter on liquid crystal polymers includes many of the newer types of system including, in the thermotropic class, those in which the elongated mesogenic side groups are linked from around their mid-points to the backbone by flexible linkages so that they lie in line with the backbone. With additional information on synthesis, this chapter is of great relevance to chemists.

Finally, as Senior Reporter and Editor, it has been a great pleasure to work with the authors of the various chapters and to unify these into a text that hopefully will be of service and interest to professional chemists who may read it either to extend their knowledge or because they work in one of the many areas upon which the expanding sphere of influence of liquid crystals now impinges.

G.W. Gray

1 Structural classification of liquid crystals

A.J. Leadbetter
Rutherford Appleton Laboratory

1.1 Introduction

The question of the classification of liquid crystal (LC) phases is one giving rise to a great deal of difficulty and confusion. The principal reason is precisely also that which makes research in LCs so fascinating, namely the extremely rich variety of phase behaviour and the rate of new discoveries over the last decade. There is now, however, enough firm understanding to make it timely to outline the basic structural classification. This will be done in as factual a manner as possible

without using arbitrary models or complex theoretical ideas. The key to the understanding of the phase behaviour we now have lies in good structural experiments coupled with the realization that the observed structures can be understood on the basis of the behaviour of one-, two-, and three-dimensionally ordered systems. This article will concentrate on structural aspects of LCs as determined by diffraction experiments, but it must be emphasized that it is essential that the systems studied are well defined and that this requires miscibility, texture and calorimetric studies in addition to direct structural measurements. Although a broad structural understanding now exists, there are many subtleties of structural behaviour and new facets are still being uncovered, so the plan of this article will be to emphasize first the fundamental structural parameters of the various phases and then separately to embellish this with a brief outline of such things as the different kinds of smectic A phase found in some highly polar compounds and the appearance of reentrant phase behaviour and modulated structures. The main emphasis will be placed on the classical thermotropic materials composed of long rod- or lathe-shaped molecules, but a brief review of the structural features of the LC phases formed from disc-shaped molecules (discotics) will also be included.

The nomenclature for the wide variety of smectic (S) phases is often a source of confusion. The use of letters as in the designation smectic A or smectic B has no fundamental significance whatsoever. It arose historically as a convenient label so that, as new phases were discovered, new letters were used and in some cases the nomenclature has changed with time. However, the current usage is now well established and almost universally accepted so that, although imperfect, this nomenclature must continue to be used.

One of the major points which has been resolved over the last few years is the question of whether a material is a *liquid* crystal or a *true* (but disordered) crystal. In principle this is very simply resolved given a careful definition of words (and good experiments), and we shall use the following nomenclature: when a structure has long-range order of the molecular positions in three dimensions it is a crystal, despite the presence of various other kinds of disorder, and structures having less than this degree of positional order, but retaining some aspects of order above that possessed by an isotropic liquid, are properly called liquid crystals. The latter are nematic or smectic and will be called N, smectic A (S_A), etc., whereas the former which have usually also been called smectics will simply be called E, G, etc., phases. These crystal phases do possess considerable disorder of molecular orientation and are certainly mesophases — intermediate between fully ordered crystal and isotropic liquid — so they are included in this chapter.

The principal phase types which have been established are summarized in Figure 1.1. These are grouped into types which it will be the main purpose of this chapter to describe and explain. The many and varied phase transitions involving the phases of Figure 1.1 have themselves been the subject of much research over the last decade and there remain many unsolved problems. They will not be considered here, but it should be noted that, in general, increase of temperature takes a compound through one or more of the phases of Figure 1.1 in the order from left to right[1]. Transitions between orthogonal and tilted phases are common, mostly involving B (or S_B^h) and S_A phases with G, S_F or S_C phases. There are also a number of intriguing, and unexplained, anomalies such as G–S_F–B transitions (crystal–liquid crystal–crystal)[2] and the D phase which is formed only on heating S_C[3].

1.2 Structural concepts

1.2.1 Introduction

The structure of an ordered crystal is very simply described using three positional parameters (plus appropriate thermal parameters) per atom. These parameters are obtained from a detailed X-ray or neutron diffraction study where the aim is usually to have up to about ten independent Bragg intensities per structural parameter. Such a structure for an ordered crystalline phase of a substance which displays several disordered crystal and LC phases is shown in Figure 1.2[4]. For isotropic liquids, on the other hand, only diffuse diffraction peaks are observed and the structure can only be described in terms of a radial distribution function which, for a molecular liquid, contains no atomic detail. The diffraction pattern of all the phases of relevance here show, because of the large amount of disorder always present, at most a very small number (of order 10) of independent Bragg reflections plus considerable structured diffuse scattering. In no case therefore is any very detailed structural description at an atomic level possible. Instead, the essential features of the structure must be defined in terms of a few simple parameters. These are concerned with:

1. positional order;
2. bond orientational (or unit lattice vector) order;
3. molecular orientation.

1.2.2 Positional order

One may consider positional order in terms of the extent to which the position of an average molecule or group of molecules shows translational symmetry. There are three possible situations, illustrated in Figure 1.3, which may be described in terms of an ideal average lattice structure modified by a correlation function[5, 6]

$$P(\mathbf{G}, \mathbf{r}) = \langle \exp i\,\mathbf{G} \cdot [\mathbf{u}(\mathbf{r}) - \mathbf{u}(o)] \rangle$$

where $\mathbf{G}$ is a reciprocal lattice vector and $\mathbf{u}(\mathbf{r})$ is a displacement at $\mathbf{r}$. For true long-range order (LRO) (a true crystal), $P(\mathbf{G}, \mathbf{r}) \sim \langle u^2 \rangle =$ constant, and the diffraction pattern consists of infinitely sharp peaks. Note, however, that the molecular distribution function may be sharp, as in a conventional ordered crystal (Figure 1.2), or soft, as in all of the structures to be considered here, where

(a) Structures based on elongated molecules

Ordered crystal

Disordered crystal				Liquid crystal		
Layer structures				Smectic		Nematic
				Based on weakly coupled ordered layers —'two-dimensional' systems	Based on one-dimensional density wave (liquid layers)	
Orthogonal: molecules parallel to **c**, perpendicular to **ab** plane		E	B	S_B^h	S_A $(S_{A_1}, S_{A_2}, S_{A_d}, S_{\tilde{A}})$	N
Tilted: molecules parallel to **c**, at angle $<\pi/2$ to **ab** plane; **b** unique (tilt) axis	**a** > **b**	H	G	S_F	S_C	
	a < **b**	K	J	S_I	$(S_{C_1}, S_{C_2}, S_{C_d}, S_{\tilde{C}})$	
Cubic structures						
D other						

Isotropic liquid

(b) Structures based on disc-shaped molecules

	Columnar structures	Nematic	
Ordered crystal	Based on columns of molecules stacked in ordered (o) or disordered (d) manner; columns form two-dimensional lattice	N_D	Isotropic liquid
	Hexagonal: D_{ho}, D_{hd} Rectangular: D_{rd}: $P2_1/a$, $P2/a$, $C2/m$ Oblique D_{obd}		

Figure 1.1 Major phase types of liquid crystalline materials.

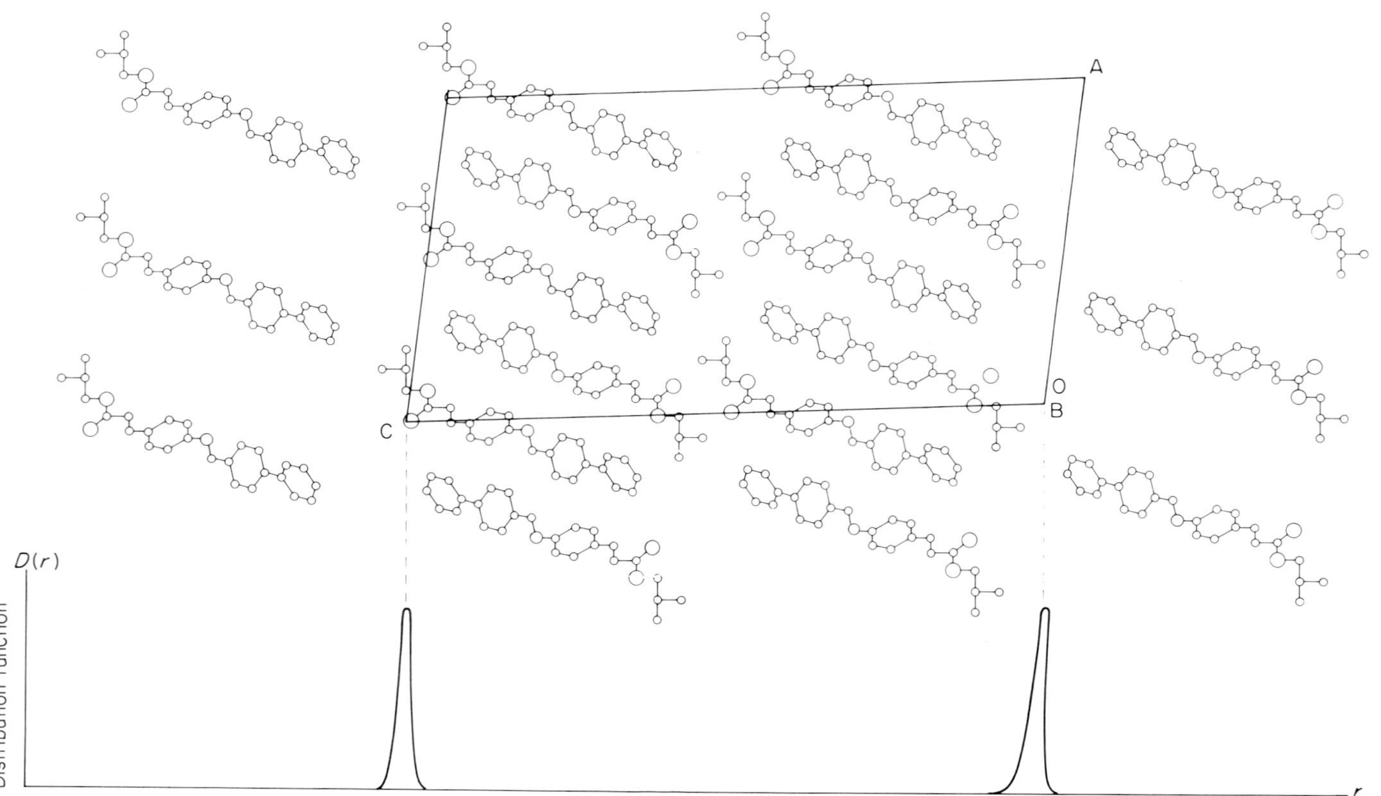

Figure 1.2 Molecular packing in the ordered crystal phase of Isobutyl-4-[4-phenylbenzylideneamino]cinnamate. This substance shows the phase behaviour: crystal–E–B–S_A–N–isotropic liquid.

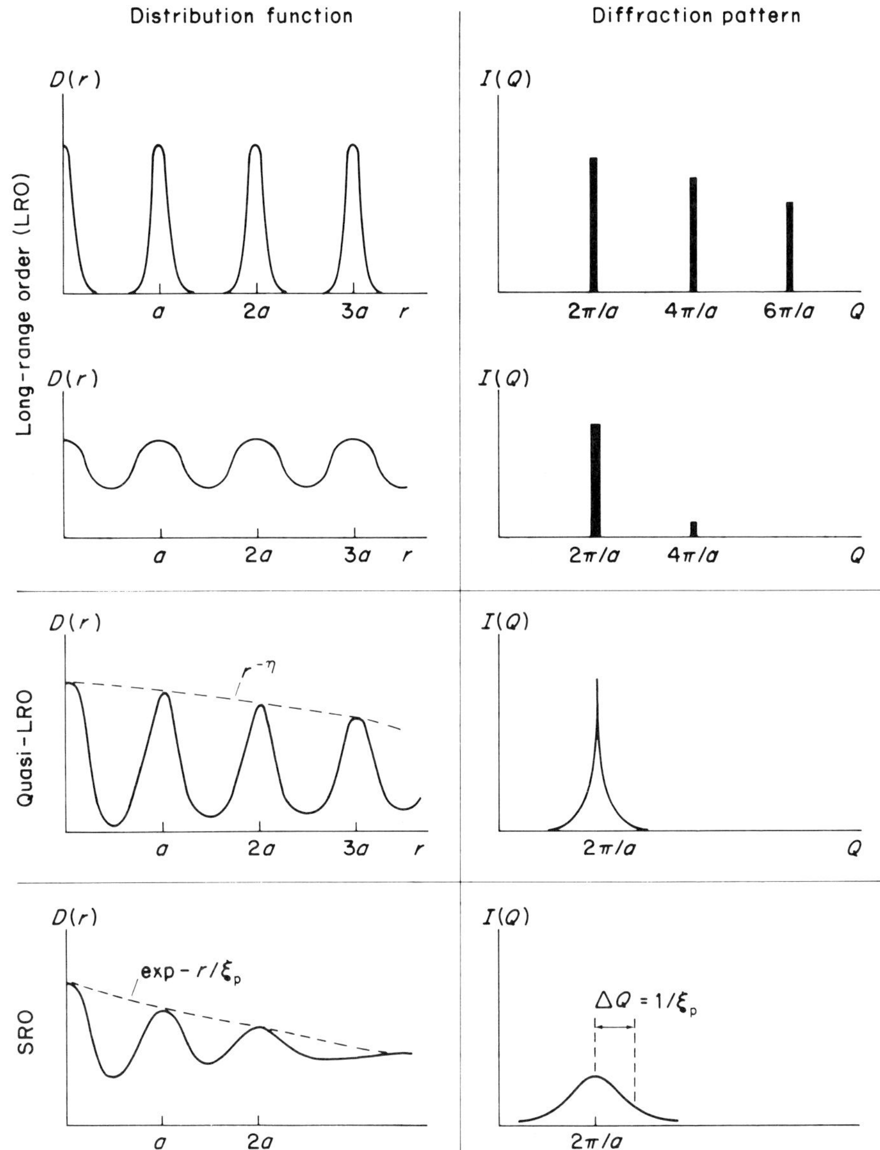

Figure 1.3 Molecular distribution functions and corresponding diffraction patterns for different kinds of positional order.

indeed it may often be a single sinusoidal function; the difference is reflected in the number and intensity of diffraction peaks which remain sharp.

The case of quasi-long-range order is very important in discussing liquid crystals and here $P(\mathbf{G}, \mathbf{r}) \sim r^{-\eta(T)}$, where $r^{-\eta}$ is a temperature-dependent quantity related to the elastic properties. This is called algebraic decay of positional order and it is revealed by diffraction peaks which are no longer delta functions but weaker singularities of the form $(\mathbf{Q} - \mathbf{G})^{\eta-2}$, where $\mathbf{Q}$ is the scattering vector.

The final case is that of short-range order (SRO) characteristic of a liquid. Here $P(\mathbf{G}, \mathbf{r}) \sim \exp - r/\zeta_p$, where ζ_p is the positional correlation length; such a structure has broad diffraction peaks of width $\Delta Q \sim \zeta_p^{-1}$.

These ideas are fundamental to understanding the structural behaviour of all the liquid and disordered crystalline phases to be described below.

1.2.3 Bond orientational order

The concept of so-called bond orientational order has emerged in recent years from theoretical considerations of two-dimensional systems[6, 7]. If one considers molecules packed in a two-dimensional array with n-fold symmetry, an order parameter may be defined as $\psi(r) = \exp i\,n\,\theta\,(r)$, where θ specifies the orientation, relative to some fixed direction, of a bond joining nearest-neighbour centres of mass. For chemists this is an unfortunate terminology but one which seems to have been accepted, but the 'bond' direction may better be thought of simply as a unit vector describing the lattice axis. The important result to emerge from theory is that bond orientational long-range order may exist without positional LRO, i.e. that the direction of the lattice axes may be conserved while the long-range periodicity of the lattice points (and molecules) is not. Formally this is expressed in terms of an orientational correlation function

$$O(r) = \langle \psi^* (\mathbf{r})\, \psi\, (o) \rangle$$

which may have three types of behaviour exactly as for $P(\mathbf{G}, \mathbf{r})$. It will be useful to list these, together with the associated positional functions, as established theoretically for two-dimensional systems for which three distinct phases are found as follows:

	$P(\mathbf{G}, \mathbf{r})$	$O(\mathbf{r})$
Two-dimensional crystal:	$r^{-\eta(T)}$ Quasi-LRO	$\langle\psi\rangle^2$ LRO
Hexatic phase:	$\exp - r/\xi_p$ SRO	$r^{-\eta_6(T)}$ Quasi-LRO
Liquid:	$\exp - r/\xi_p$ SRO	$\exp - r/\xi_o$ SRO

These results are illustrated schematically in Figure 1.4, and in fact it is easy to see qualitatively that the positional LRO may be destroyed while preserving the orientational LRO by the presence of a sufficient concentration of dislocations.

These concepts will be applied later to rationalize the structural characteristics

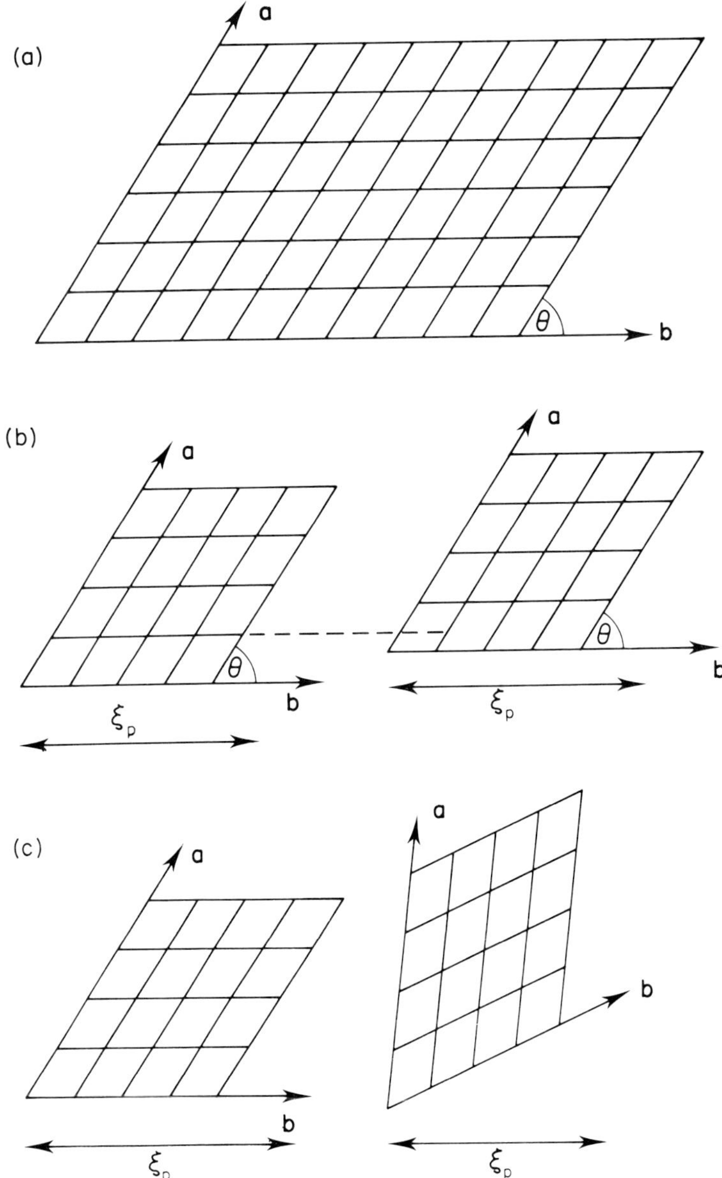

Figure 1.4 Illustration of positional and bond orientational order: (a) long-range positional and bond orientational order, (b) short-range positional order, long-range bond orientational order, (c) short-range order of both position and bond orientation. ζ_p denotes the positional correlation length.

of the 'two-dimensional' phases S_B^h, S_F, and S_I and their relationship to the more and less ordered phases existing at lower and higher temperatures respectively.

1.2.4 Molecular orientation

All of the phases of Figure 1.1(a) are characterized by a tendency of the molecules to align with their long axes more or less parallel. We consider here other aspects of molecular orientation.

It has now been clearly established by a variety of techniques[8], and especially by incoherent quasi-elastic neutron scattering[9, 10], that in all of the phases in Figure 1.1(a) the molecules rotate about their long axes with a correlation time of $\gtrsim 10^{-10}$ s. This means that any site symmetry, e.g. the sixfold axis in crystal B phases, must arise from dynamic disorder, and for the liquid-like packing in S_A and N phases, any deviation from isotropy of the molecular orientational distribution function about the long axes must be extremely small, if not zero. The existence of ferroelectric S_C^* phases shows that this distribution function is not isotropic in the S_C phase (see Section 1.6). Dielectric permittivity and relaxation measurements also strongly suggest that rotation about the short axes (end-over-end rotation) also occurs in all these phases, albeit with much longer time constants ($\gtrsim 10^{-5}$ s)[11, 12].

In general, therefore, we also expect head-to-tail disorder, and this is consistent with the structural data on the most ordered phases such as the E phase[13] and must in general be true *a fortiori* for the less ordered phases. However, in the last few years a number of examples have emerged in S_A and S_C phases of highly polar compounds where there appears an additional ordering associated with correlations in the polarity of the orientations of the molecular long axes[14, 15]. This will be discussed later.

1.3 Nematics

The nematic (N) phase is an anisotropic liquid[16]. The molecular positions show only short-range order in all directions, but the molecules do possess a quasi-long-range orientational order of their long axes which tend to be parallel to a common axis called the director **n** (Figure 1.5), and **n** is indistinguishable from −**n**. A small orienting field (e.g. 0.02*T*) is sufficient to give true long-range order of director orientation. The point group symmetry of the N phase is $D_{\infty h}$. The molecular long axis is operationally defined by the particular experiment and this can result in order parameters $[\langle P_2(\cos\theta)\rangle]$, where θ is the angle between the 'long-axis' and the director, being experiment dependent. The nature of the distribution function for the long axes is shown in Figure 1.6[5, 17].

The diffraction pattern is anisotropic but diffuse. It is always characterized by strong arcs centred at a scattering vector $Q_\perp \sim 2\pi/d$, where d is the average molecule width, and usually (but not always) by diffuse peaks at $Q_\parallel \sim 2\pi/l$, where l is the molecular length. These arise from short-range structural

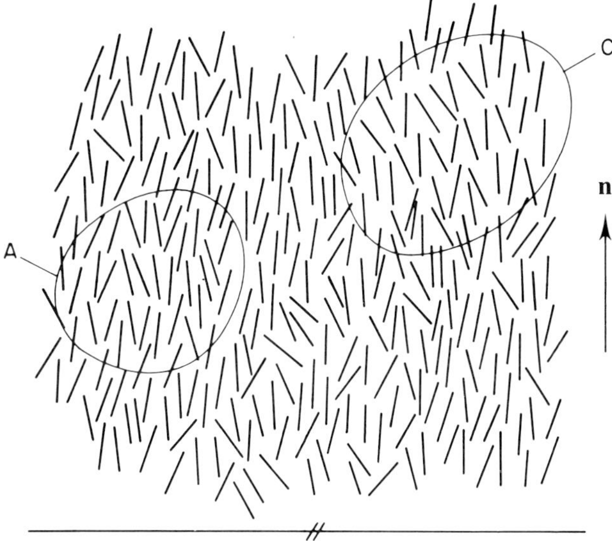

Figure 1.5 Schematic representation of nematic structures. Regions A and C represent local fluctuations of S_A- and S_C-type structural correlations.

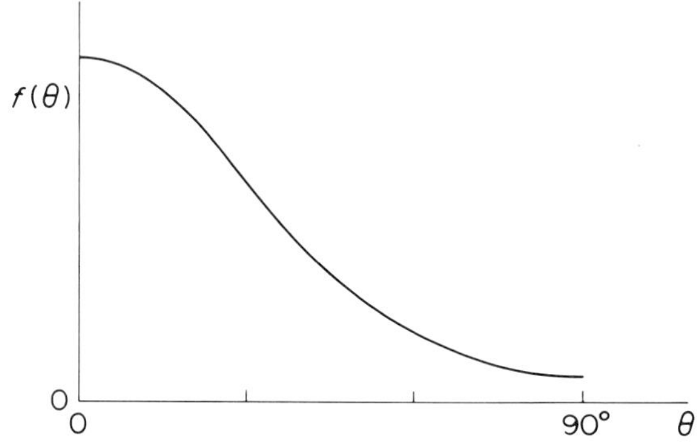

Figure 1.6 Distribution function for the molecular long axes in a nematic phase, as determined by X-ray diffraction.

correlations similar to those of smectic A or C phases, i.e. from evanescent density waves parallel (S_A) or at an angle (S_C) to the director (Figure 1.5).

When strong S_C-like fluctuations occur in the nematic, these are often known as cybotactic groups[18]. The positional correlation lengths in nematics usually correspond to a few molecular dimensions, but along the evanescent density wave they may grow rapidly, reaching several hundred molecular lengths, as the transition to a S_A or S_C phase is approached[19].

Many nematics composed of molecules with cyano end groups (e.g. the alkylcyanobiphenyls) show a strong tendency to molecular association which results in the presence of a damped density wave with a wavelength considerably longer than the molecular length (typically $\sim 1.4l$ or l + the length of an alkyl tail)[20, 21]. Examples have also been found where there exist two independent damped density waves of wavelength l and $\sim 1.4l$[22]. Many cases are now known in which two (or more) different N phases occur, separated by S_A (or S_C) phases. This reentrant behaviour is often associated with the tendency of polar molecules to associate [23] and will be discussed briefly in Section 1.4.

1.4 Smectics based on one-dimensional density waves

1.4.1 Conventional smectic A and C phases

Smectic A and C phases have long-range orientational order of the molecular long axes similar to, but with a distribution function sharper than, that for nematics. The distinguishing feature of these phases, however, is the presence of a single density wave, with a wave vector Q_A or Q_C parallel to the director for S_A phases and at an angle to it of θ_t (the tilt angle) for S_C phases[16, 23]. Usually this density wave is almost purely sinusoidal (see Figure 1.3), in contrast to the rather sharp gaussian function characteristic of an ordered crystal. Thus, although the wavefronts may properly be called layers, the usual representation of these as well-defined sharp layers is very misleading (see Figure 1.7).

In the plane perpendicular to Q_A or Q_C, the molecular positions have only short-range order with correlation lengths of a few molecular diameters. The presence of a single density wave involving long-range positional order would give a one-dimensional crystal. However, it has long been known that fluctuations in such a one-dimensional system result in a logarithmic divergence of the mean square displacement according to the ratio of sample dimension (D) and molecular length (l), i.e. $\langle u^2 \rangle = C(T) \ln (D/l)$[16, 23]. This means that the positional correlation function has algebraic decay ($P(r) \sim r^{-\eta(T)}$). This has been confirmed[24, 25] by very high resolution X-ray diffraction measurements which show that the diffraction pattern does not show the infinitely sharp delta function peaks $I(Q) = \delta(\mathbf{Q} - \mathbf{G})$ (Bragg reflections) of the true crystal, but rather peaks of the form $I(Q) \sim (\mathbf{Q} - \mathbf{G})^{\eta - 2}$. In fact these are so sharp that in conventional medium resolution experiments they are indistinguishable from ordinary Bragg reflections.

The layer spacing d in S_A phases is normally close to but somewhat shorter than the molecular length, and the same is true of the quantity $d \cos \theta_t$ for S_C phases. The S_A phase has point symmetry D_∞ and is optically uniaxial while the S_C phase has C_{2h} symmetry and is optically biaxial.

1.4.2 S_A and S_C phases of highly polar molecules

In many cases of pure compounds or mixtures of molecules containing strongly polar groups[26], an additional ordering appears which is based on a tendency to antiferroelectric ordering of molecules, i.e. to an additional ordering of

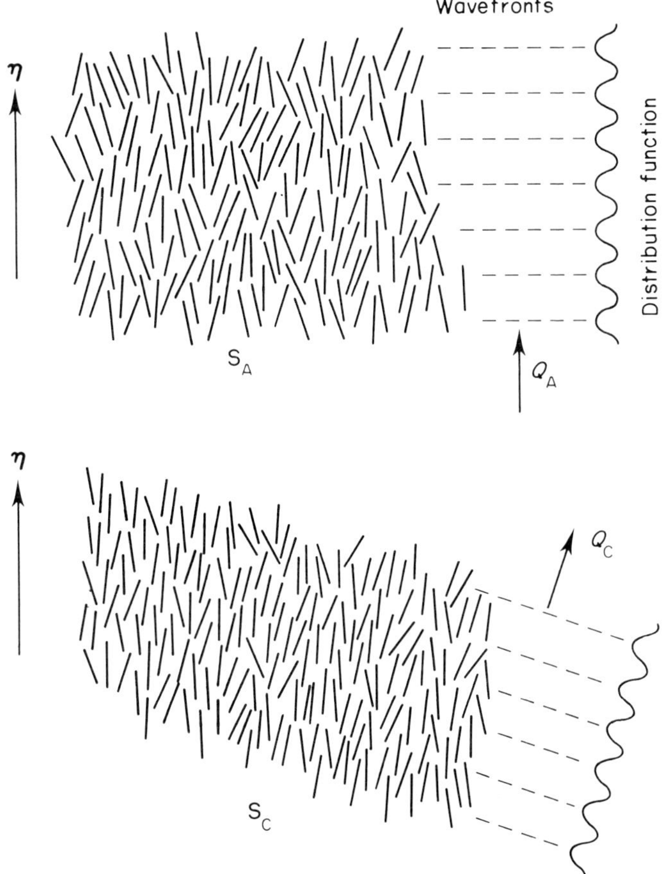

Figure 1.7 Schematic drawing of S_A and S_C structures.

orientations of the molecular long axes[14, 15, 23, 27, 28]. The known structures are illustrated in Figure 1.8 and may be described as follows:

- S_{A_1}, S_{C_1} Conventional phases with completely random head-to-tail disorder
- S_{A_2}, S_{C_2} Bilayer phases with antiferroelectric ordering of the molecules
- S_{A_d}, S_{C_d} Semi-bilayers with partial molecular association
- $S_{\tilde{A}}$, $S_{\tilde{C}}$ Modulated antiferroelectric ordering within the layers to give a 'ribbon' structure, called 'antiphase' structures

More than one of these phases may exist in a given pure compound (or mixture) and they are then often associated with reentrant behaviour. This involves unusual phase sequences as a function of temperature, such as: S_{A_1}–N–S_{A_d}–N, which is illustrated by the phase diagram shown in Figure 1.9. A convincing explanation of this complex phase behaviour has been suggested[8, 23, 27] in

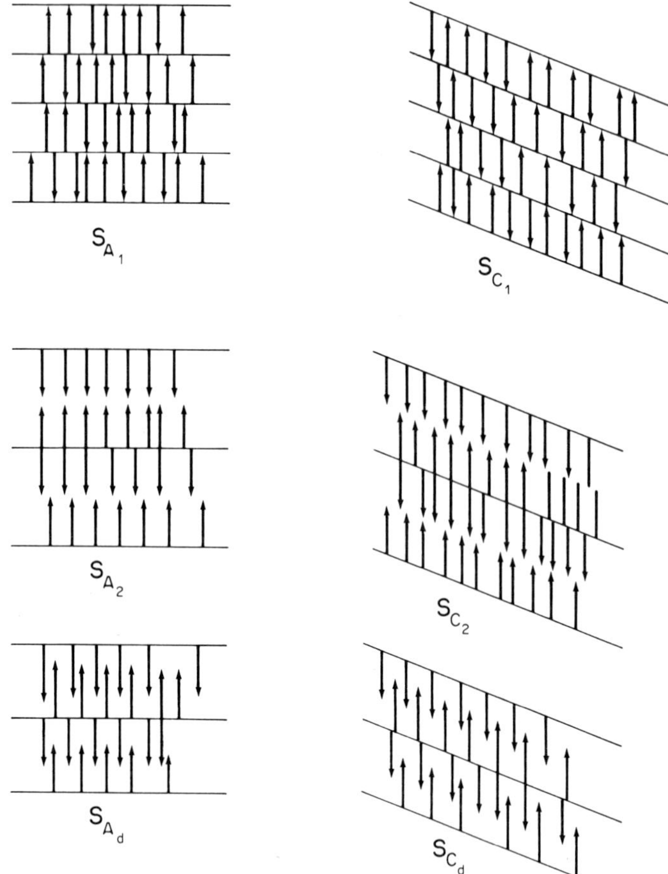
S_{A_1}
S_{C_1}
S_{A_2}
S_{C_2}
S_{A_d}
S_{C_d}

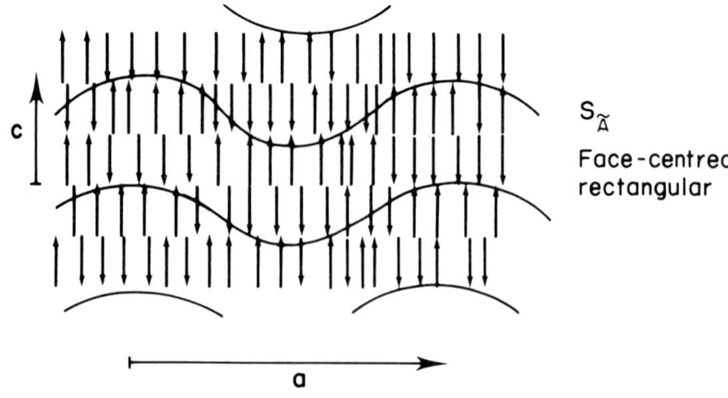
c
a
$S_{\tilde{A}}$
Face-centred
rectangular

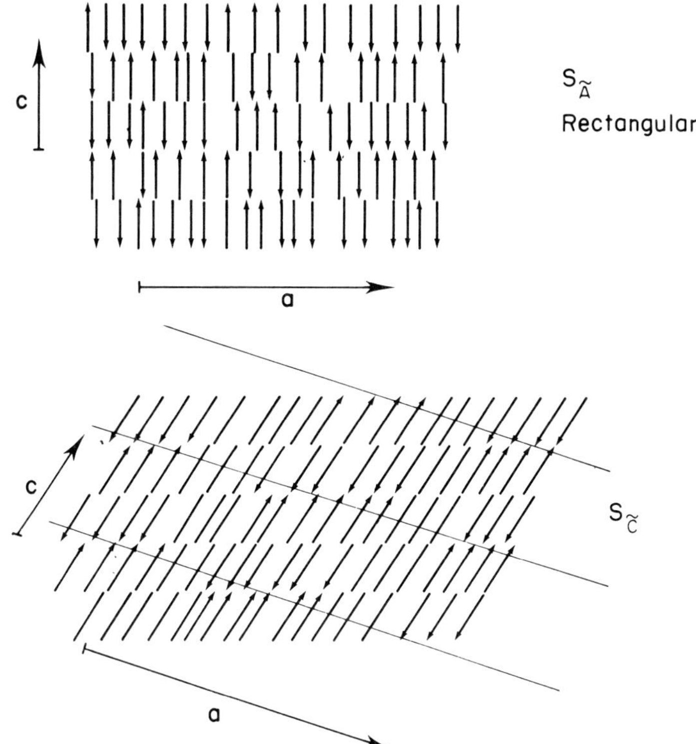

Figure 1.8 Schematic drawing of $S_{\tilde{A}}$ and $S_{\tilde{C}}$ structures found in polar compounds. The axes of the two-dimensional lattice associated with the molecular polarization in the $S_{\tilde{A}}$ and $S_{\tilde{C}}$ structures are designated by **a** and **c**.

terms of competition between the tendency to condensation of a density wave of wavelength comparable to the molecular length l and an antiferroelectric association of molecules to give a characteristic pair length l', with l and l' incommensurate. A case has recently been reported[29] where the above competition resolves itself in the coexistence of two colinear incommensurate density waves of type S_{A_2} and S_{A_d}.

The antiphase structures are extremely interesting because, while they still consist of liquid-like layers associated with a one-dimensional wave, there is additionally a two-dimensional molecular polarization structure, as shown in Figure 1.8. The structures are described as ribbon-like in that in the third dimension (normal to the paper) there is no modulation or periodicity.

1.5 Smectics based on weakly coupled two-dimensional layers: S_B^h, S_F, S_I

The structures of these phases are based on weakly coupled ordered layers[30, 31]. The layers are soft, in that normally only one or two orders of

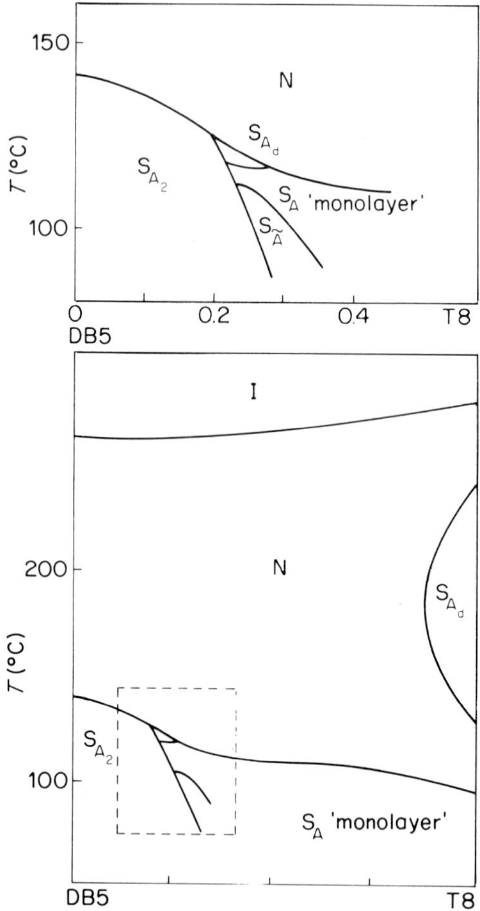

Figure 1.9 Phase diagram of $C_5H_{11}C_6H_4OOCC_6H_4OOCC_6H_4CN$ (DB5) and $C_8H_{17}OC_6H_4OOCC_6H_4CH{=}NC_6H_4CN$ (T8) showing reentrant phase behaviour.

diffraction from the layers are observed, showing that they are best described as almost sinusoidal density waves (very similar to the S_A and S_C phases).

There is positional LRO (or quasi-LRO; no definitive very high resolution measurements have been reported) along the layer-normal. Normal to the layer wave vector, i.e. within the layers, the molecules are packed with what is formally short-range positional order, but the correlation length is of order 100 Å or more, which is about an order of magnitude longer than in S_A or S_C phases. The main additional — and novel — feature is the existence of true LRO of bond orientation in three dimensions. This is revealed by the existence of a well-defined symmetry in the diffraction pattern of a bulk sample[32–34]. This means that although there is only short-range order of the in-plane (*xy*) molecular positions and that this also extends only over one or two layers, despite the long-range order of the *z*

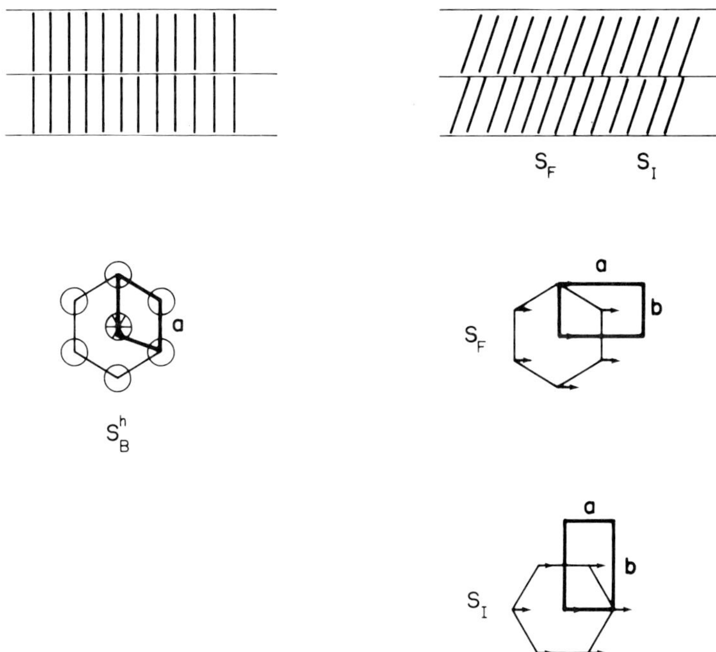

Figure 1.10 Schematic drawing of the local structures of S_B^h, S_F and S_I phases. The arrows denote the tilt direction of the molecules.

coordinate, the layers may not rotate relative to each other. The orientation of the local lattice axes has long-range order both within and between layers.

In all three phases (Figure 1.10), the molecules are packed in a hexagonal (S_B^h) or quasi-hexagonal (S_F, S_I) manner. The S_B^h phase is uniaxial with true hexagonal symmetry, which reflects a dynamic average over six-fold molecular orientational disorder with a rotational correlation time of $\sim 10^{-11}$ s. The S_F and S_I phases are tilted, with monoclinic C_{2h} symmetry. Adopting the usual convention of labelling the unique (tilt) axis as **b**, then in S_F, $\mathbf{a} > \mathbf{b}$ and the tilt is towards the side of the quasi-hexagon, while for S_I, $\mathbf{a} < \mathbf{b}$ and the tilt is towards an apex; both lattices are *C*-centred monoclinic.

The existence of these phases can be rationalized in a very cogent way[6] by taking into account the effect of weak interlayer coupling on the theoretical predictions for true two-dimensional systems, as outlined in Section 1.2. Thus, noting that algebraic decay of correlation functions is accompanied by an infinite corresponding susceptibility so that quasi-LRO in the true two-dimensional case is converted to true LRO in the real, weakly coupled, liquid crystal situation, the effect in the three two-dimensional based phases is given in Table 1.1. The S_B^h, S_F and S_I phases are thus often known as stacked hexatic phases, but it must be emphasized that the bond orientational LRO now extends over all three dimensions.

Table 1.1 Relation of weakly coupled layer phases to the theoretical behaviour of two-dimensional systems

Correlation function	True two-dimensional system		Weakly coupled two-dimensional system—liquid crystal	
$P(\mathbf{G}, \mathbf{r})$ $O(\mathbf{r})$	$r^{-\eta(T)}$ $\langle\psi\rangle^2$	Two-dimensional crystal	$\langle u^2\rangle$ $\langle\psi\rangle^2$	True crystal E, J, etc.; see Section 1.7
$P(\mathbf{G}, \mathbf{r})$ $O(\mathbf{r})$	$\exp - r/\xi_P$ $r^{-\eta_6(T)}$	Hexatic phase	$\exp - r/\xi_P$ $\langle\psi\rangle^2$	Stacked hexatic phases (S_B^h, S_F, S_I)
$P(\mathbf{G}, \mathbf{r})$ $O(\mathbf{r})$	$\exp - r/\xi_P$ $\exp - r/\xi_O$	Liquid	$\exp - r/\xi_P$ $\exp - r/\xi_O$	Liquid layers (S_A, S_C)

There has been a suggestion that S_I differs from S_F in addition to the different tilt direction by having algebraic decay of positional order within the layers[34]. The experiments were, however, made on a powder sample and have not been confirmed on other oriented samples[35].

1.6 Chiral compounds

1.6.1 Cholesterics and chiral smectics

If a chiral molecule forms a nematic or tilted smectic liquid crystalline phase or if it may be incorporated in appropriate concentration into such a phase, then the structure becomes chiral and contains a helical twist[16]. This does not happen for the S_A or S_B^h phases.

The helical N structure is called a cholesteric (Ch) phase and possesses a helical twist of the director with a pitch (Figure 1.11) of the order of the wavelength of light (a few thousand ångströms) which results in Bragg scattering of light and characteristic colour effects. This pitch is much longer than the molecular length, and it can be removed by the action of rather small electrical or magnetic fields which align the molecules along the field.

The tilted S phases all have point group symmetries of C_{2h}, but when the molecule is chiral, this drops to a C_2 symmetry[36], the C_2 axis being perpendicular to the molecular tilt direction and to the layer-normal (Figure 1.11). Two results then follow: firstly, there is a macroscopic polarization along the C_2 axis so that the phases are ferroelectric, and secondly there is a helical precession of the tilt axis along the direction perpendicular to the layers (again of pitch $> 10^3$ Å) so that in the absence of an external field the phases are helielectric. The helix is unwound by an electric field along C_2 and the polarization can thus be switched between two equivalent but opposite orientations (Figure 1.11). This effect may well form the basis of new fast ($\sim$ μs) optoelectronic display devices[37].

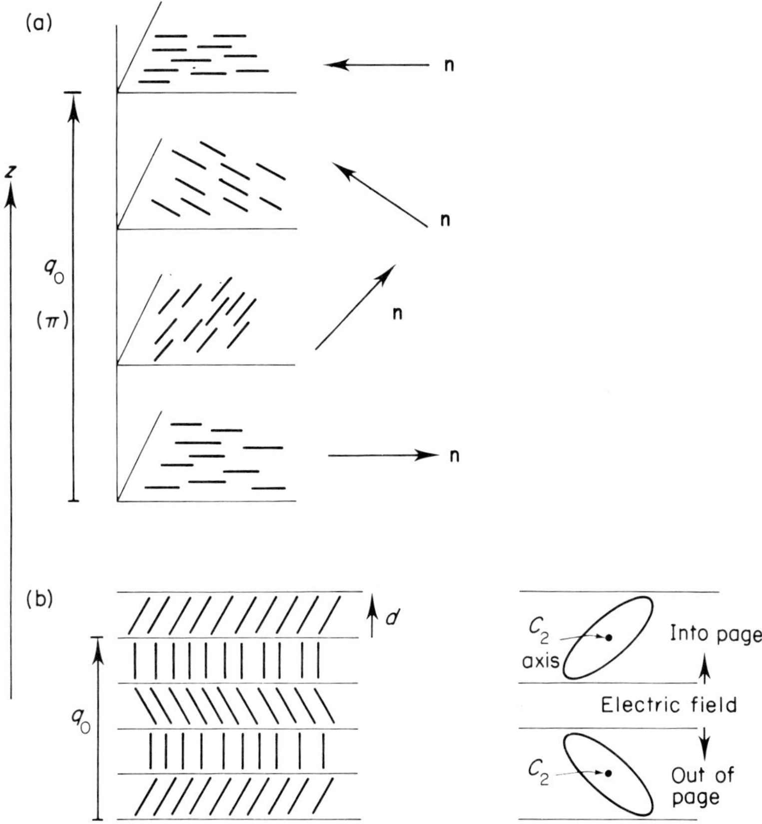

Figure 1.11 Chiral mesophases (q_0 denotes the helix pitch and z the helix axis): (a) cholesteric, (b) helical S^*_C, S^*_F and S^*_I.

1.6.2 Blue phases

These phases, named for their visual appearance, often appear in a narrow range of temperature (typically less than 0.5 °C) between the Ch and isotropic liquid phases for Ch phases of relatively short pitch (<3000 Å). They have been extensively studied over the last few years, and it now appears that there are at least three such phases[1, 38–40]. The phases are optically isotropic, exhibit Bragg (light) scattering, and monocrystals may be grown showing distinct crystal faces. Such experiments have been used to examine the structure of the phases and it now seems to be well established that blue phase I (BPI) is body-centred cubic ($I4_132$) and BPII is primitive cubic ($P4_232$); the lattice parameters are the pitch lengths. BPIII is also known as 'blue fog' and no structural symmetry has been observed.

Extensive theoretical studies have also been made and have shown that the cubic symmetries are obtained via the packing of helical structures[41, 42]. It is at least conceivable that despite the difference in scale, there are similarities between these blue cubic structures and those of the D phases discussed in Section 1.7.

1.7 Disordered crystals

1.7.1 Layer structure

At least six different basic phase types of layer structure having three-dimensional positional LRO of the molecules (and hence also necessarily LR bond orientational order) are known[1, 3, 30, 43]. They are listed in Figure 1.1(a). These phases are true crystals [44, 45] characterized by considerable orientational disorder of the molecules[34, 46], and although they are indeed strongly layer-like in structure and are still often referred to as smectics, this name is better reserved for phases lacking positional LRO in at least one dimension. Moreover, once it is realized that they are crystals, it is also apparent that many more layer-like structures are possible, so it is not surprising that a number of variants on the six basic structures have been found recently and more will undoubtedly emerge. The common feature of orientational molecular disorder (which may include disorder of internal molecular configuration) is revealed by diffraction spots which fall off extremely rapidly in intensity with scattering vector, so that only a very small number of independent reflections is observed. For example, it is not common to observe more than five orders of layer reflection, and the appearance of only a first-order plus an extremely weak second-order reflection is very common. The disorder is also revealed by large entropy and volume changes at the transition from one of these disordered phases to the ordered crystal phase found at lower temperatures[47]. The magnitude of these changes is typically about an order of magnitude larger than for any of the phase transitions at higher temperatures.

The structures of the six basic phase types are very simply related to those of the stacked hexatic phases, as shown in Figure 1.12. The crystal B, G, and J phases are derived from S_B^h, S_F and S_I by the development of positional LRO within the layers — and hence, necessarily, between the layers. The simple B phase is uniaxial with true hexagonal symmetry associated with dynamic rotational disorder about the six-fold axis. The G and J phases have tilted quasi-hexagonal packing of the molecules, the tilts being respectively towards a side and an apex of the hexagon. The crystallographic unit cells are *C*-centred monoclinic, with $\mathbf{a} > \mathbf{b}$ for G and $\mathbf{a} < \mathbf{b}$ for J.

The E, H and K phases are related to the B, G and J phases by a loss of molecular rotational freedom, which results in the adoption of a herringbone instead of a (quasi)-hexagonal packing and the loss of the *C*-centring of the unit cell. The E phase thus has orthorhombic symmetry[48, 49] and the H or K phases monoclinic symmetry[43, 50] (Figure 1.12). In all these 'herringbone' phases, the molecules remain disordered with respect to rotations of π about both long and short molecular axes[10, 13].

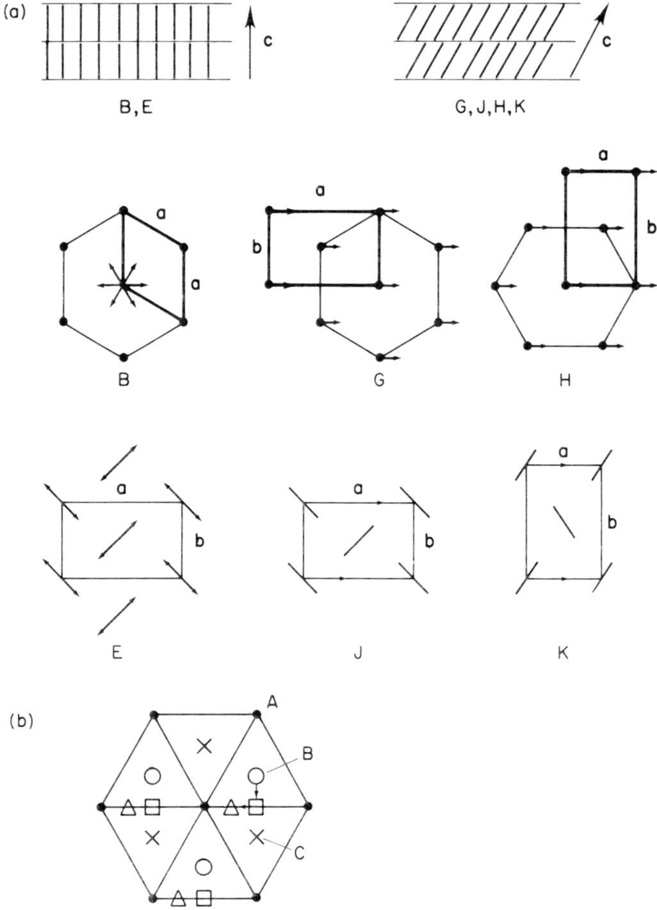

Figure 1.12 (a) Structures of disordered crystal phases B, E, G, H, J and K. For the orthogonal phases the disorder about the long molecular axis is indicated by the arrows while for the tilted phases the single arrows denote the tilt direction. (b) Layer stackings of hexagonal layers in different B-type phases. A, B and C denote the highest symmetry positions of molecules in adjacent layers, while the open squares and triangles show the lower symmetry positions which have been observed[52].

A number of variations on the above structures are now known. Thus the hexagonal layers in B-type structures may be stacked to give moni-, bi- or trilayer unit cells[51], i.e. in configurations AAA..., ABAB..., ABCA (Figure 1.12) with layers B, C not necessarily in the trigonal positions relative to A, so that the crystal symmetry may be lower than hexagonal[52]. In many cases, two or three of these structures occur with change of temperature in the same compound, with no detectable thermal effects, despite the fact that a change of symmetry may be involved. Similarly E-type orthorhombic phases with both monolayer[48, 49] and bilayer[4, 13] unit cells have been identified.

A particularly interesting modification of B phases is the existence of ripple-like modulations of the layers[51, 53]. The amplitude and wavelength of the modulations may be temperature dependent, and they appear to be related both to the occurrence of stacking transitions[52] and to the B to G (orthogonal to tilted) structural transition[54].

1.7.2 Cubic structures

Two apparently different types of cubic mesophase are known to occur in thermotropic liquid crystalline systems[1, 3]. The first [55] is known to occur in only four compounds of the form

C_nH_{2n+1}—O—(benzene ring bearing X)—(benzene ring)—COOH

with n = 16 or 18
X = NO_2 or CN

It has often been called a smectic D phase, but as it is a crystal we shall call it simply a D phase.

The phase behaviour for the (16, NO_2) compound is S_C–D–S_A–isotropic liquid, while for the other compounds it is similar, except that the S_A phase does not occur. The phase is optically isotropic and has recently been shown[56] by single-crystal diffraction experiments to be a three-dimensional crystal with a primitive cubic structure. The lattice parameters are of order 90 Å, compared with monomer molecular lengths of approximately 35 Å, and there are about 800 molecules in the unit cell. Diffuse X-ray scattering shows that locally molecules are packed with positional SRO as in the S_C layers, but the detailed nature of the structure which gives rise to the cubic LRO is not yet established. It is likely, however, to involve bundles of molecules packed into quasi-spherical or rod-like agglomerations which can pack together with the appropriate cubic symmetry.

The second type of cubic structure[57] is formed by compounds in the series

C_nH_{2n+1}O—(benzene ring)—CONH—HNOC—(benzene ring)—OC_nH_{2n+1}

with $n \geqslant 8$

The cubic phase occurs between an ordered crystal and an S_C phase and also appears to have a primitive cubic space group with a unit cell size about half that for the D phases.

Much interesting work remains to be done in elucidating the nature of these phases and their relation to the adjacent smectic C phases.

1.8 Discotic liquid crystals

1.8.1 Introduction

The first discotic liquid crystal was synthesized and identified as recently as 1977[58] and a reasonably good understanding of the structures of the various phases now exists[59–61]. These phases are formed from molecules having more or less flat aromatic cores with usually six, but sometimes four, lateral substituents, normally alkoxy or ester (or more complex) groups, with at least five

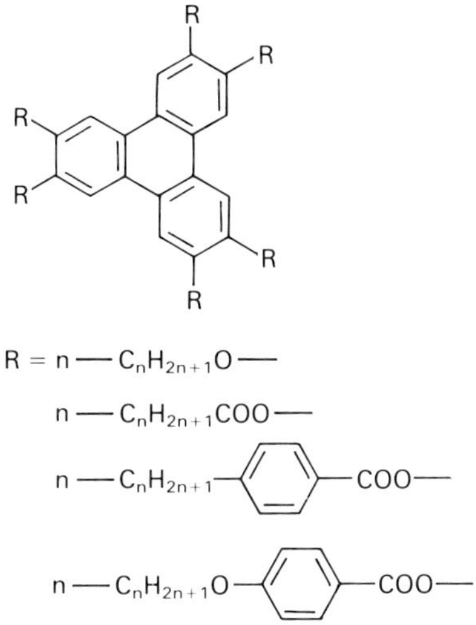

R R R R R R

$R = n—C_nH_{2n+1}O—$

$n—C_nH_{2n+1}COO—$

$n—C_nH_{2n+1}—C_6H_4—COO—$

$n—C_nH_{2n+1}O—C_6H_4—COO—$

Figure 1.13 Typical discotic molecules.

carbon atoms (Figure 1.13). The structures are based on the tendency of the molecular discs to align with their short axes (the normals to the average molecular planes) parallel. The steric driving force for this is as intuitively obvious as it is in the case of elongated molecules discussed above. However, the presence of the disordered side chains is crucial to the formation of discotic mesophases. The particular molecular disc-like shape determines the basic types of liquid crystal formed and is used in the nomenclature that describes the phases. These are of two fundamental types: nematic and columnar.

1.8.2 Nematics (N_D)

As in conventional nematics, these phases are anisotropic fluids with a single order parameter associated with the tendency of the discs to align parallel. The disc-normals thus tend to point along a common direction — the director **n** (Figure 1.14).

Reentrant behaviour has also been observed in discotic materials[62], with an N_D phase appearing, with change of temperature, both above and below a

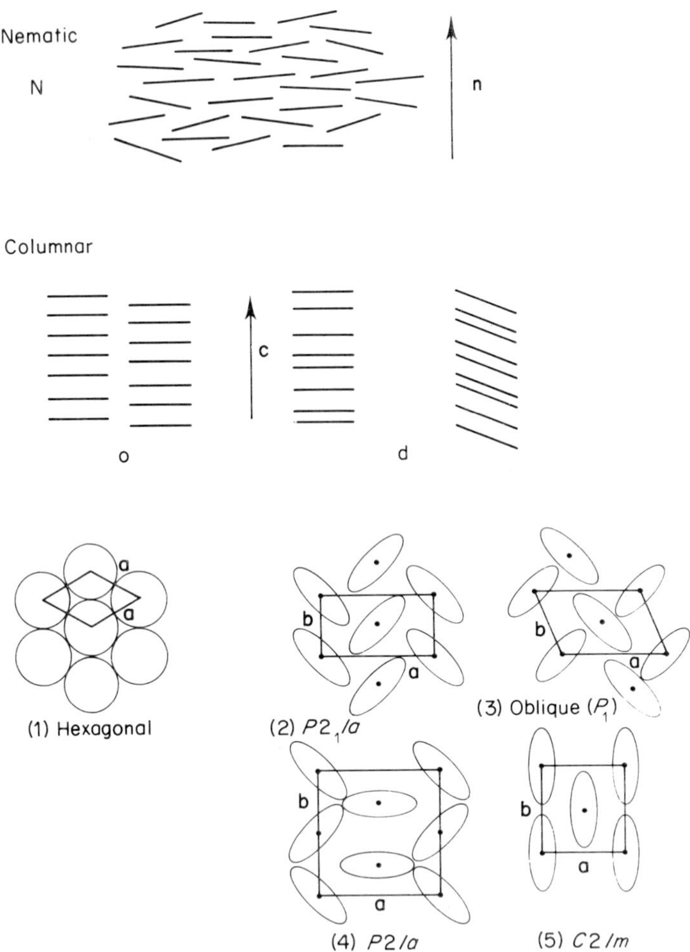

Figure 1.14 Structures of discotic phases. The lines represent the planes of the molecules. (1)–(5) are plan views of the two-dimensional lattices of the columns. Circles denote that the molecular planes are normal to the column axis and ovals that they are tilted.

columnar phase. The difference in the two N phases is not clear, but may be connected with molecular association as for rod-like reentrant behaviour.

1.8.3 Columnar phases

These phases are characterised by stacked columns of molecules, the columns being packed together to form a two-dimensional crystalline array. Within the columns, the molecular positions may have short-or long-range order, but the columns are not in register along their axes, so these phases are not three-dimensionally crystalline. No high-resolution X-ray work has been done, and in only a few cases have single-domain samples been obtained; no general statements can therefore be made about the true nature or extent of order in the ordered columns or about the correlation lengths along the disordered columns. The appearance, however, of several orders of sharp spots on X-ray photographs showing the two-dimensional columnar packing[60] strongly suggests that this two-dimensional array has true positional LRO. The columnar phases can thus be described as having two-dimensional positional LRO and three-dimensional bond orientational LRO.

All of the phases so far known can be characterized by the order (o) or disorder (d) of the molecular stacking in the columns (Figure 1.14) and by the two-dimensional lattice symmetry of the columnar packing. They are all labelled by D to denote discotic mesophases and the known phases are summarized below and illustrated in Figure 1.14:

D_{ho}, D_{hd}	Hexagonal, ordered and disordered
D_{rd}	Rectangular disordered
$P2_1/a$	
$P2/a$	Plane space group symmetries
$C2/m$	
D_{obd}	Oblique disordered

Optical observations have shown that the average molecular plane is not normal to the column axis for all the rectangular phases and, although the tilt angles are not accurately known, the relative orientation of tilt vectors in neighbouring columns is included in the space group designation.

Note finally that chiral versions of some of these phases also exist[59, 63], the N_D phase being clearly analagous to the conventional cholesteric phase.

1.9 Conclusions

The brief outline presented above of the principal structural features of the known mesophases of rod- and disc-shaped molecules shows that a good descriptive understanding of the structures and of their interrelations now exists. Furthermore, an underlying theoretical understanding of the existence of the principal phase types based on the behaviour of low-dimensional systems also exists, together with a rationalization of the existence of the various different types of N, S_A and S_C phases provided by the competition between density wave and antiferroelectric ordering. However, we are still very far from a detailed understanding at a molecular level of why particular phases form, why some are tilted and some not, and about what drives many of the phase transitions. Indeed, the whole subject of phase transitions in mesogens and their relation to modern theoretical work has been a very rich field of study and will continue to be so in the future.

Finally, it may be remarked that molecular shape is a dominant factor in determining the fundamental type of mesophase formed, as, for example, between smectic and columnar, although anomalies like the cubic phases still remain unexplained. The recent discovery of new mesophase structures intermediate between those of columnar and lamellar phases, based on molecules with a rod-like core and half-discs and each having three chains containing seven or more carbon atoms at each end, is therefore highly significant[64]. This shows that our growing understanding is leading to new discoveries and that the field will remain alive for a long time yet.

1.10 References

1. D. Demus, S. Diele, S. Grande and H. Sackmann, *Adv. Liq. Cryst.*, **6**, 1 (1983).
2. J.W. Goodby, G.W. Gray, A.J. Leadbetter and M.A. Mazid, *J. Phys. (Paris)*, **41**, 591 (1980).
3. G.W. Gray and J.W. Goodby, *Smectic Liquid Crystals*, Leonard Hill, Glasgow, Scotland (1984).
4. A.J. Leadbetter, M.A. Mazid and K.M.A. Malik, *Mol. Cryst. Liq. Cryst.*, **61**, 39 (1980).
5. A.J. Leadbetter in *The Molecular Physics of Liquid Crystals*, ed. by G.R. Luckhurst and G.W. Gray, Academic Press, London (1979), Ch. 8, p. 284.
6. R. Birgeneau and J.D. Litster, *J. Phys. (Paris) Lett.*, **39**, L-399 (1978).
7. D.R. Nelson and B.I. Halperin *Phys. Rev.*, **B19**, 2457 (1979).
8. S. Chandrasekhar and N.V. Madhusudana, *Proc. Indian Acad. Sci. (Chem. Sci.)*, **94**, 139 (1985).
9. A.J. Leadbetter and R.M. Richardson in *The Molecular Physics of Liquid Crystals*, ed. by G.R. Luckhurst and G.W. Gray, Academic Press, London (1979), Ch. 20, p. 451.
10. R.M. Richardson, A.J. Leadbetter and J.C. Frost, *Mol. Phys.*, **45**, 1163 (1982).
11. H. Kresse, *Adv. Liq. Cryst.* **6**, 109 (1983).
12. L. Bata and A. Buka, *Mol. Cryst. Liq. Cryst.*, **63**, 307 (1981).
13. A.J. Leadbetter, J.C. Frost, J.P. Gaughan and M.A. Mazid, *J. Phys. (Paris)*, **40**, C3–185 (1979).
14. F. Hardouin, A.M. Levelut, M.F. Achard and G. Sigaud, *J. Chim. Phys.*, **80**, 53 (1983).
15. N.V. Madhusudana, *Proc. Indian Acad. Sci. (Chem. Sci.)*, **92**, 509 (1983).
16. P.G. de Gennes, *The Physics of Liquid Crystals*, Oxford University Press (1974).
17. A.J. Leadbetter and E.K. Norris, *Mol. Phys.*, **38**, 669 (1979).
18. A. de Vries, *Mol. Cryst. Liq. Cryst.*, **10**, 219 (1970).
19. J.D. Litster, *Philos. Trans. R. Soc. London, A*, **309**, 145 (1983).
20. A.J. Leadbetter, R.M. Richardson and C.N. Colling, *J. Phys. (Paris)*, **36**, C1–37 (1975).
21. A.J. Leadbetter, J.C. Frost, J.P. Gaughan, G.W. Gray and A. Mosley, *J. Phys. (Paris)*, **40**, 375 (1979).
22. G.J. Brownsey and A.J. Leadbetter, *Phys. Rev. Lett.*, **44**, 1608 (1980).
23. J. Prost, *Adv. Phys.*, **33**, 1 (1984).
24. J. Als-Nielsen, J.D. Litster, R.J. Birgeneau, M. Kaplan, C.R. Safinya, A. Lindegaard-Andersen and B. Mathiesen, *Phys. Rev.*, **B22**, 312 (1980).
25. J. Als-Nielsen in *Symmetries and Broken Symmetries*, ed. by N. Bocarra, IDSET, Paris (1981), p. 107.
26. N.H. Tinh, *J. Chim. Phys.*, **80**, 83 (1983).
27. J. Prost and P. Barois, *J. Chim. Phys.*, **80**, 65 (1983).
28. A.M. Levelut, *J. Phys. (Paris) Lett.*, **45**, L-603 (1984).
29. B.R. Ratna, R. Shashidhar and V.N. Raja, *Phys. Rev. Lett.*, **55**, 1476 (1985).
30. P.A.C. Gane, A.J. Leadbetter and P.G. Wrighton, *Mol. Cryst. Liq. Cryst.*, **66**, 247 (1981).
31. J.J. Benattar, J. Doucet and M. Lambert, *Phys. Rev.*, **A20**, 2505 (1979).
32. A.J. Leadbetter, J.P. Gaughan, B.A. Kelly, G.W. Gray and J.W. Goodby, *J. Phys. (Paris)*, **40**, C3–178 (1979).
33. R. Pindak, D.E. Moncton, S.C. Davey and J.W. Goodby, *Phys. Rev. Lett.*, **46**, 1135 (1981).
34. J.J. Benattar, F. Moussa and M. Lambert, *J. Chim. Phys.*, **80**, 99 (1983).
35. J. Budai, R. Pindak, S.C. Davey and J.W. Goodby, *J. Phys. (Paris) Lett.*, **45**, 1053 (1984).
36. R.B. Meyer, L. Liebert, L. Strzelecki and P. Keller, *J. Phys. (Paris) Lett.*, **36**, L69 (1975).
37. N.A. Clark and S.T. Lagerwall, *Appl. Phys. Lett.*, **36**, 899 (1980).
38. R. Barbet-Massin, P.E. Cladis and P. Pieranski, *Phys. Rev.*, **A30**, 1161 (1984).
39. V.A. Kizel and V.V. Prokhorov, *J.E.T.P.*, **60**, 257 (1984).
40. Th. Blumel and H. Stegemeyer, *Z. Naturforsch.*, **40a**, 260 (1985).
41. H. Grebel, R. Hornreich and S. Shtrikman, *Phys. Rev.*, **A28**, 1114 (1983).

42. H. Grebel, R. Hornreich and S. Shtrikman, *Phys. Rev.*, **A30**, 3264 (1984).
43. P.A.C. Gane, A.J. Leadbetter, P.G. Wrighton, J.W. Goodby, G.W. Gray and A.R. Tajbakhsh, *Mol. Cryst. Liq. Cryst.*, **100**, 67 (1983).
44. D.E. Moncton and R. Pindak, *Phys. Rev. Lett.*, **43**, 701 (1979).
45. P.S. Pershan, G. Aeppli, J.D. Litster and R.J. Birgeneau, *Mol. Cryst. Liq. Cryst.*, **67**, 205 (1981).
46. J.J. Benattar, A.M. Levelut, L. Liebert and F. Moussa, *J. Phys. (Paris)*, **40**, C3–115 (1979).
47. A. Wiegeleben, L. Richter, J. Deresch and D. Demus, *Mol. Cryst. Liq. Cryst.*, **59**, 329 (1980).
48. S. Diele, P. Brand and H. Sackmann, *Mol. Cryst. Liq. Cryst.*, **17**, 163 (1972).
49. J. Doucet, A.M. Levelut, M. Lambert, L. Liebert and L. Strzelecki, *J. Phys. (Paris)*, **36**, C1 (1975).
50. J. Doucet, A.M. Levelut and M. Lambert, *Phys. Rev. Lett.*, **32**, 301 (1974).
51. A.J. Leadbetter, M.A. Mazid and R.M. Richardson in *Liquid Crystals*, ed. by S. Chandrasekhar, Heyden and Son, London (1980), p. 65.
52. J. Collett, L.B. Sorensen, P.S. Pershan, J.D. Litster, R.J. Birgeneau and J. Als-Nielsen, *Phys. Rev. Lett.*, **49**, 553 (1982).
53. A.J. Leadbetter, M.A. Mazid, B.A. Kelly, J. W. Goodby and G.W. Gray, *Phys. Rev. Lett.*, **43**, 630 (1979).
54. P.A.C. Gane and A.J. Leadbetter, *J. Phys. C.*, **16**, 2059 (1983).
55. D. Demus, G. Kunicke, J. Neelson and H. Sackmann, *Z. Naturforsch.*, **23a**, 84 (1968).
56. G.E. Etherington, A.J. Leadbetter, X.J. Wang, G.W. Gray and A.R. Tajbakhsh, *Liquid Crystals*, **1**, 209 (1986).
57. D. Demus, A. Gloza, H. Hartung, I. Rapthel and A. Wiegeleben, *Cryst. Res. Technol.*, **16**, 1445 (1981).
58. S Chandrasekhar, B.K. Sadashiva and K.A. Suresh, *Pramana*, **9**, 471 (1977).
59. S. Chandrasekhar, *Philos. Trans. R. Soc. London, A*, **309**, 93 (1983).
60. A.M. Levelut, *J. Chim. Phys.*, **80**, 149 (1983).
61. C. Destrade, P. Foucher, H. Gasparoux and N.H. Tinh, *Mol. Cryst. Liq. Cryst.*, **106**, 121 (1984).
62. N.H. Tinh, P. Foucher, C. Destrade, A.M. Levelut and J. Malthete, *Mol. Cryst. Liq. Cryst.*, **111**, 277 (1984).
63. A.M. Levelut, P. Oswald, A. Ghanem and J. Malthete, *J. Phys. (Paris)*, **45**, 745 (1984).
64. J. Malthete, A.M. Levelut and N.H. Tinh, *J. Phys. (Paris) Lett.*, **46**, L-875 (1985).

2 Liquid crystal behaviour in relation to molecular structure

K.J. Toyne
University of Hull

2.1 Introduction

The phenomenon of liquid crystallinity was discovered almost a hundred years ago for cholesteryl benzoate[1] and since that time many thousands of mesogens have been prepared and catalogues listing their transition temperatures have been published[2, 3]. Liquid crystals were initially of interest as examples of a new state of matter and until about 1965 most research work involved the preparation of homologous series of compounds with a particular core structure and a study of how the transition temperatures and types of mesophase changed within that series. The types of mesophase encountered at that time were limited mainly to nematic and smectic A, B or C phases, but from 1965 onwards, with the expectation and eventual realization of effective electrooptical display devices using liquid crystals, research in the area has increased significantly and all three main classes of mesophase are now used in technological applications[4, 5] e.g., nematic (N) (for the twisted-nematic device, see Chapter 3), smectic (S) (particularly S_A for laser-addressed and electrically-addressed displays, S_C^* for ferroelectric devices, see Chapter 4) and cholesteric (Ch) (for use as thermochromic materials, see Chapter 5, and for phase-change displays). The demand for new materials has meant that many novel, potentially useful, systems have been prepared, but frequently the behaviour thoughout an homologous series has not been examined in detail, as was formerly the case.

In this chapter the main features of the effect of structure on liquid crystal behaviour will be discussed. Some of the firmly established relationships will be presented briefly and greater detail can be found in earlier reviews or in more general books on liquid crystals[6–12]. It is hoped, however, that some impression of recent major aspects of liquid crystal research can be conveyed without giving too much attention to interesting but currently peripheral topics. Since comprehensive coverage of all aspects of liquid crystals embraced by the title of this chapter was not possible in the space available, certain major omissions were necessary, such as a discussion of discotic materials and of how molecular structure influences the polymorphic types of S phase in smectogens. Several publications on the former topic include references [13–19] and Chapter 1, and references [20–22] give detailed accounts of S phases.

Although Ch materials (chiral nematics) are not overtly covered in this chapter, most of the generalizations made for the effects of molecular structure on N behaviour are also applicable to cholesterics. The influence of molecular structure on the types of S phase has not been so successfully rationalized, and the importance of achieving such an understanding is particularly acute in the search for acceptable S_C phases for use in ferroelectric devices (see Chapter 4 and also a chapter in reference [20]). A further difficulty in attempting to rationalize molecular structures and S phases is that all too frequently such phases are reported without being uniquely identified.

2.2 General aspects of core length: terminal, linking and lateral groups

Most of the compounds listed by Demus *et al.*[2, 3] and Kelker *et al.*[23] are aromatic systems, probably because aromatic units give mesophases of good thermal stability[24] and because, in general, such compounds are relatively easy to prepare. In recent years non-aromatic compounds have become much more common[25] and it has been shown that they also may generate mesophases of high thermal stability (see page 32; Reinitzer's cholesteryl benzoate is mainly alicyclic). All such mesogens (excluding plastic crystal materials and discogens) have a characteristic geometric molecular shape (described as being rod-like or lath-like) which gives rise to anisotropic intermolecular forces with the interactions between the ends of the molecules being weaker than the lateral associations. In general, therefore, any structural feature (see compound **1**), such as increased core length (e.g. n changed from 1 to 2), terminal groups (A or C) or linking groups (B) which give reasonable rigidity to an almost linear molecule should be conducive to liquid crystal formation; even small

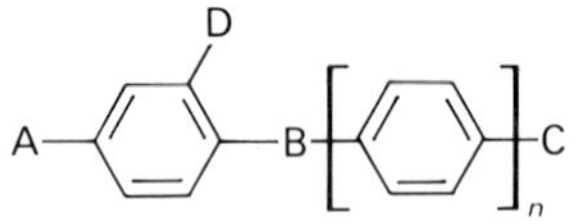

1

lateral groups (D) may often be tolerated, whereas larger lateral groups would be expected to decrease the broadside association of molecules and lead to a reduction in mesophase thermal stability (see page 47). Any terminal group (A or C) which extends the molecule along the molecular axis without increasing the molecular breadth too much increases the thermal stability of a mesophase. It is also true that such substituents increase the anisotropy of molecular polarizability ($\Delta\alpha$), particularly if they conjugate with the aromatic ring, and the theory of Maier and Saupe[26–28] suggests a link between the N–isotropic liquid transition temperature (T_{N-I}) and $\Delta\alpha$. The terminal group efficiency order which has been compiled for N and Ch phases in aromatic systems is[8]

$$\mathrm{Ph} > \mathrm{NHCOCH_3} > \mathrm{CN} > \mathrm{OCH_3} > \mathrm{NO_2} > \mathrm{Cl} > \mathrm{Br} > \mathrm{N(CH_3)_2} > \mathrm{CH_3} > \mathrm{F} > \mathrm{H}$$

A common terminal substituent is the n-alkyl or n-alkoxy group and by plotting the transition temperatures (mesophase-to-mesophase or mesophase-to-isotropic liquid) against the number of carbon atoms in the alkyl (C_nH_{2n+1}) or alkoxy ($C_{n-1}H_{2n-1}O$) group for an homologous series, one of several possible smooth curves is revealed. As the series is ascended, the values for the transition temperature alternate (usually members with odd n have the higher and members with even n have the lower values, and for alkoxy compounds, the oxygen atom is equivalent to a CH_2 group), and the alternation effect becomes less pronounced. The shape of the curve varies and the lines for the odd or even series may, for example, both fall gradually, both rise gradually or rapidly, one may fall and one may rise, both may fall initially to minima and then rise again, etc.; examples of two important, structurally related systems which give quite different shapes are shown in Figures 2.1 and 2.2 (these curves are discussed further on page 32 and illustrate how misleading conclusions may be drawn when comparing different series by choosing single compounds with a certain alkyl chain). The alternation in simple cases is thought to be related to the change in $\Delta\alpha$ as the chain length increases and, assuming a rigid, extended zig-zag alkyl chain, the odd series have more carbon–carbon bonds along the molecular axis giving the higher $\Delta\alpha$[29]; although a single extended *s-trans* conformation is unlikely, it is believed to be a major preferred conformation[30–32]. With longer alkyl chains, the differences between the two curves will decrease as the flexibility of the chain increases and, even for initially rising curves, the T_{N-I} values should eventually fall as the increasing chain flexibility begins to disrupt the N ordering. However, the expected decrease in the T_{N-I} value may not be detected if stable S phases appear.

Another common terminal group is the cyano group, and molecules with this substituent show the additional feature of antiparallel correlation of molecular dipoles (or antiferroelectric short-range order)[33, 34] which results in higher clearing points than with comparable non-polar substituents (e.g. alkyl, see later) and affects other physical properties; for example, terminally cyano-substituted mesogens have high values for the ratio of k_{33}/k_{11}[35, 36].

Suitable linking groups (B in compound **1**), which preserve the linearity of the molecule and by being unsaturated also extend conjugation between two rings are $-(CH{=}CH)_n-$, $-C{\equiv}C-$, $-CH{=}N-$, $-CH{=}\underset{\substack{\downarrow \\ O}}{N}-$, $-N{=}N-$, $-N{=}\underset{\substack{\downarrow \\ O}}{N}-$, $-C(=O)O-$, $-CH{=}CH-C(=O)O-$, whereas $-C(=O)O-H$ (or $-CH{=}CH-C(=O)O-H$), which is formally a terminal group, forms cyclic dimers and therefore functions as a linking group. Linking groups such as $-CH_2CH_2-$ and $-CH_2O-$ which do not permit conjugation from one ring to the next do not enhance the molecular rigidity and are poor linking groups between aromatic rings but may be acceptable in alicyclic systems (see page 44).

The effect of lateral substituents[37] is discussed in more detail later (page 47), particularly for fluoro and cyano substituents and for large lateral substituents. Apart from those cases where the broadening effect of the substituent is minimized by being structurally shielded, lateral substituents lower clearing points by increasing the polarizability across the molecular axis (so reducing $\Delta\alpha$) and by increasing the intermolecular separation. The decrease in T_{N-I} is related to the broadening effect of the substituent and dipolar effects are insignificant. The effect of lateral substituents on the thermal stability of S phases is different and for a strongly dipolar substituent there is a smaller decrease in T_{S-N} or T_{S-I} compared to a less dipolar substituent of the same size, which suggests that dipolar association is a more important interaction in smectics than in nematics (see page 49).

Osman has recently discussed the nature of the core[38], terminal substituents[39] and lateral substituents[40] in influencing the thermodynamic stabilities of mesophases, and some of these issues will be discussed later.

2.3 Terminal cyano compounds

The primary impetus to the commercial exploitation of liquid crystals for electrooptic display devices came in 1972 when Gray *et al.*[41] prepared

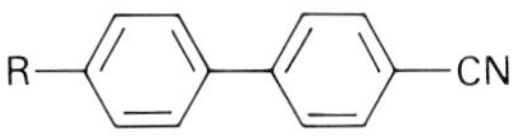

R–C6H4–C6H4–CN

K series

2

RO–C6H4–C6H4–CN

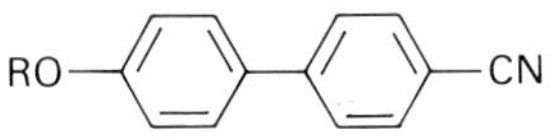

M series

3

4-n-alkyl- and 4-n-alkoxy-4′-cyanobiphenyls (**2** and **3** respectively) and their terphenyl analogues to meet the requirements of materials for the twisted-nematic device[42]. The development of the ideas leading to the selection of these colourless compounds[7, 43] which have the required properties of room temperature N ranges, chemical and photochemical stability, high resistivity and large positive dielectric anisotropy ($\Delta\varepsilon$ *ca.* +11) and the background to the cooperation of several workers in testing and utilizing these compounds provides an interesting record[44] of the origins of commercial liquid crystal activity in Britain. The transition temperatures for some alkyl- and alkoxy-cyanobiphenyls (**2** and **3**) are given in Table 2.1 and the values for the K series are plotted in Figure 2.1. These show that in both series S properties (S_A) appear at about the same chain length and the alkoxy compounds have significantly higher clearing points than the corresponding alkyl compounds (see the terminal group efficiency order page 30). The higher clearing points associated with the alkoxy group can be ascribed to the greater conjugation provided by the lone pairs of electrons on the oxygen atom, and for similar reasons we would also expect 4-alkylamino-4′-cyanobiphenyls[46–48] to have higher $T_{N\text{–}I}$ values (e.g. 4-butylamino-4′-cyanobiphenyl has a monotropic $T_{N\text{–}I}$ at 101.7 °C; the alkylamino compounds also have higher m.p.s. and higher $\Delta\varepsilon$ of about +27[46]), whereas an electron-attracting group such as perfluoroalkyl would oppose the polarity of the cyano group and probably give a reduced thermal stability[49]. Similarly, a comparison of 4-perfluoroalkyl-4′-alkylaminobiphenyls[49] and 4,4′-bis(alkylamino)biphenyls[50] shows that the former compounds have the higher clearing points.

In 1972 *trans*-4-alkylcyclohexanecarboxylic acids[51] were reported to be nematogenic, showing that even simple compounds without a conjugated aromatic core can be mesogens, and the cyclohexane unit was used in other systems to give, for example, 4-X-phenyl *trans*-4-alkylcyclohexanecarboxylates where X = CN[52], OR, R[53, 54], etc. It is worth noting that the first liquid crystal studied was a cholesterol derivative, largely composed of cyclohexane units[1], and that both cyclohexane and bicyclo(2.2.2)octane derivatives had been considered by Dewar and coworkers[55–58], although the compounds prepared did not reveal the true usefulness of these units for generating compounds with high $T_{N\text{–}I}$ values.

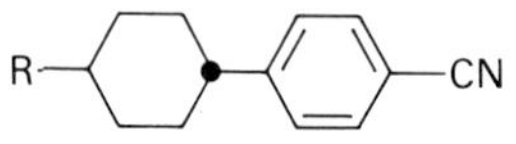

R–(cyclohexyl)–(phenyl)–CN
PCH series
4 [59]

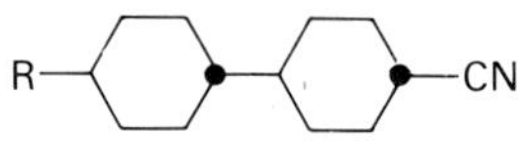

R–(cyclohexyl)–(cyclohexyl)–CN
CCH series
5 [60]

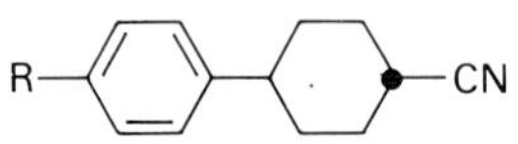

R–(phenyl)–(cyclohexyl)–CN
6 [61]

Each and both of the phenyl rings in compound **2** were replaced by *trans*-1,4-cyclohexane to give compound **4** (PCH series), compound **5** (CCH series) and compound **6**, and an even closer link with the steroid skeleton was provided by the perhydrophenanthrene **7**, which has a $T_{N\text{–}I}$ of 91 °C[62, 63].

The PCH and CCH series gave extremely useful nematogens (see Table 2.2; Figure 2.2 shows a plot of the values for the PCH series); for example, PCH5[64] has a $\Delta\varepsilon$ of +9.9, a viscosity at 20 °C of 21.0 cP* and a birefringence

*1.0 P = 1.0 g/(cm s) = 1.0 St × *d*.

Table 2.1 Transition temperatures (°C) for 4-n-alkyl-4′-cyanobiphenyls (**2**)[a] and 4-n-alkoxy-4′-cyanobiphenyls (**3**)[a][45]

Compounds 2								Compounds 3							
R	C		S_A		N		I	RO	C		S_A		N		I
CH_3	.	109	—		[.	45]	.								
C_2H_5	.	75	—		[.	22]	.	CH_3O	.	105	—		(.	85.5)	.
C_3H_7	.	66	—		(.	25.5)	.	C_2H_5O	.	102	—		(.	90.5)	.
C_4H_9	.	48	—		(.	16.5)	.	C_3H_7O	.	74.5	—		(.	64)	.
C_5H_{11}	.	24	—		.	35	.	C_4H_9O	.	78	—		(.	75.5)	.
C_6H_{13}	.	14.5	—		.	29	.	$C_5H_{11}O$	.	48	—		.	68	.
C_7H_{15}	.	30	—		.	43	.	$C_6H_{13}O$	.	57	—		.	75.5	.
C_8H_{17}	.	21.5	.	33.5	.	40.5	.	$C_7H_{15}O$	.	54	—		.	74	.
C_9H_{19}	.	42	.	48	.	49.5	.	$C_8H_{17}O$	.	54.5	.	67	.	80	.
$C_{10}H_{21}$	.	44	.	50.5	—		.	$C_9H_{19}O$	.	64	.	77.5	.	80	.
$C_{11}H_{23}$	.	53	.	57.5	—		.	$C_{10}H_{21}O$	.	59.5	.	84	—		.
$C_{12}H_{25}$	.	48	.	58.5	—		.	$C_{11}H_{23}O$	.	71.5	.	87.5	—		.
								$C_{12}H_{25}O$	.	70	.	90	—		.

[ ] Virtual transition.
( ) Monotropic transition.
[a]Members of the K and M series are indicated by a numerical suffix which is three times the number of the carbon atoms in the terminal chain, e.g. K15 is the pentyl compound and M12 is the butoxy compound.

NC— —C_6H_{13}

7

(Δn) of +0.12 to +0.13 and, although $\Delta\varepsilon$ is lower than for the K series, the lower viscosity gives faster response times and the lower Δn is advantageous in cholesteric–nematic phase-change displays. When the second phenyl ring is replaced by cyclohexane, the purely alicyclic CCHs are produced; these also give enantiotropic N phases with relatively high clearing points. Predictably, these compounds show an even lower $\Delta\varepsilon$ (+3 to +4)[68] and have $\Delta n < 0.1$; they were the first class of nematogens shown to have *negative* diamagnetic anisotropy and are useful for producing high-contrast guest–host and cholesteric–nematic phase-change displays[67]. Another unusual feature of the CCHs is that their S thermal stability *decreases* with increasing length of alkyl chain; CCH3, with propyl and cyano-groups of similar size, appears to give the most efficient S packing[69], whereas CCH7 shows only a nematic phase. Clearly the relationship between $\Delta\alpha$ and the clearing point of a liquid crystal as proposed by Maier and Saupe[26–28] is not valid in these cases because $\Delta\alpha$ decreases in the order K series, PCH series and CCH series[70].

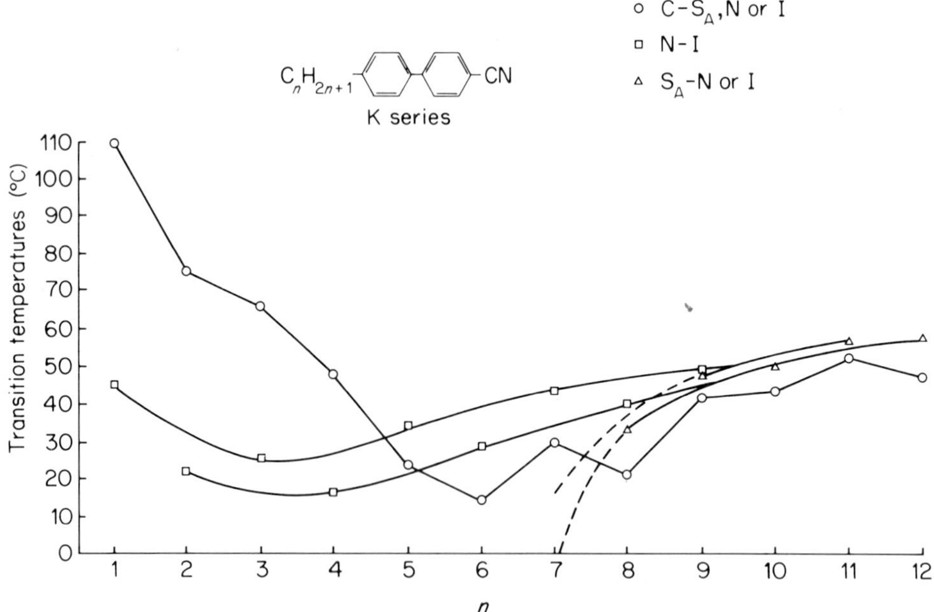

Figure 2.1 Transition temperatures (°C) for compounds of the K series.

Table 2.2 Transition temperatures (°C) for 4-(*trans*-4-n-alkylcyclohexyl)benzonitriles (**4**)[59, 64–66] and *trans*-4-(*trans*-4-n-alkylcyclohexyl)cyclohexylcarbonitriles (**5**)[60, 67]

Compounds **4**; PCH

R	C		S_A		N		I
CH_3	.	38	—		(.	−25)	.
C_2H_5	.	40	—		(.	4)	.
C_3H_7	.	36	—		.	46	.
C_4H_9	.	41	—		.	41	.
C_5H_{11}	.	31	—		.	55	.
C_6H_{13}	.	42	—		.	47	.
C_7H_{15}	.	30	(.	17)	.	59	.
C_8H_{17}	.	33	—		.	54	.
C_9H_{19}	.	35	—		.	57	.
$C_{10}H_{21}$	.	49	(.	31.5)	.	57.5	.
$C_{12}H_{25}$	.	48	—		.	60	.

Compounds **5**; CCH

R	C		S		S		S		S_B		N		I
C_2H_5	.	29	—		—		—		.	46	.	49	.
C_3H_7	.	58	(.	18	.	44	.	48	.	57)	.	80	.
C_4H_9	.	28	—		—		—		.	54	.	79	.
C_5H_{11}	.	62	—		—		(.	43	.	52)	.	85	.
C_7H_{15}	.	71	—		—		—		—		.	83	.

( ) Monotropic transition.

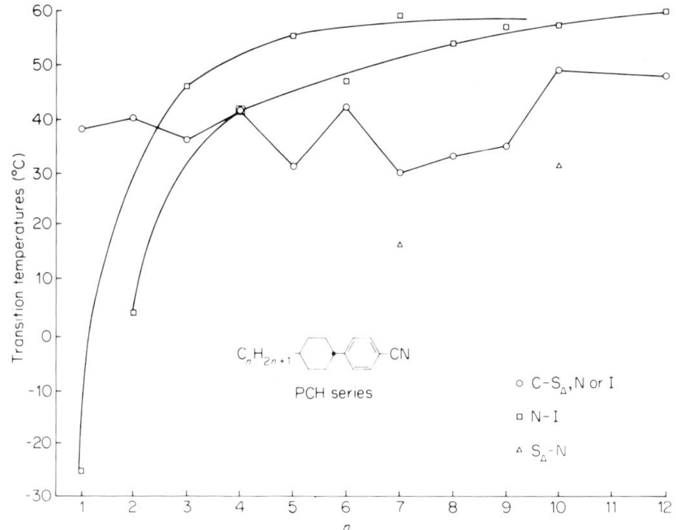

Figure 2.2 Transition temperatures (°C) for compounds of the PCH series.

In the light of these two examples it is surprising that much lower clearing points were obtained for compounds **6** and this kind of result has led to the useful idea that compounds of highest T_{N-I} are obtained when the molecule has one high and one low polarizability region[7], whereas alternating low, high, low, ... polarizability regions give poor N stability. An alternative way of expressing the situation is to consider whether or not there is a separation of planar (sp^2 or sp carbon atoms) and tetrahedral regions (sp^3 carbon atoms) in a molecule. This point is elaborated further on page 39, but Figure 2.3 illustrates how separating the cyano group from an aromatic ring reduces T_{N-I} but a cyano group in a completely saturated system gives high T_{N-I} compounds. The figure also illustrates another general observation that the bicyclo(2.2.2)octane system is consistently better than the cyclohexane or benzene ring in generating compounds with high T_{N-I} (see page 41).

When it was realized that non-aromatic hydrocarbon units such as cyclohexane and bicyclo(2.2.2)octane, when appropriately placed in cyano compounds, gave mesophases of high thermal stability, then even 'simpler' non-aromatic

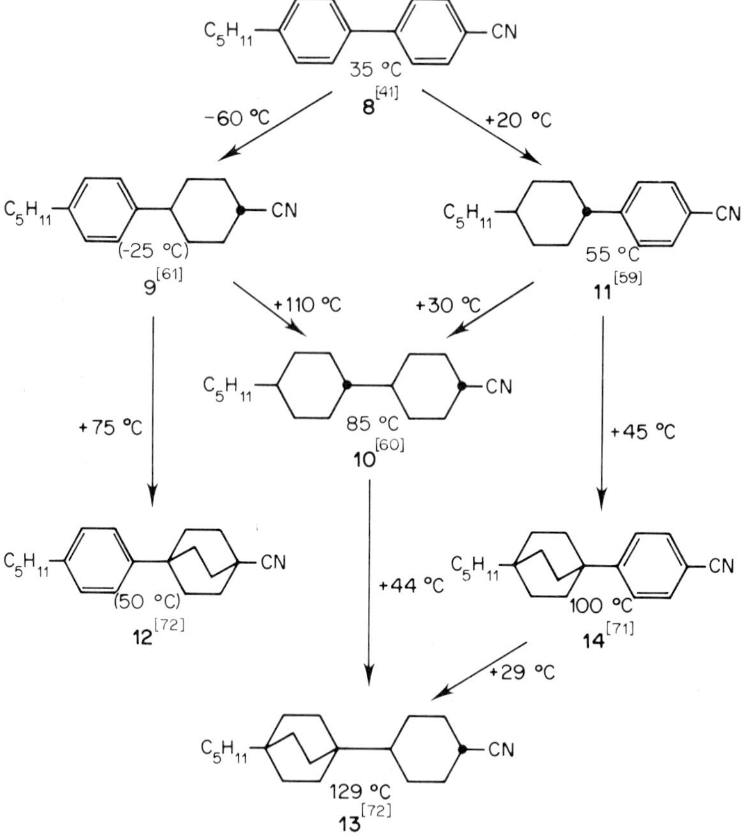

Figure 2.3 T_{N-I} values (°C) for several terminal cyano compounds[73].

hydrocarbons without polar groups were prepared to see if they were mesogenic (see page 38).

A variety of heterocyclic compounds related to the 4-n-alkyl-4′-cyanobiphenyls (**2**) have also been reported. None of these compounds shows an enantiotropic

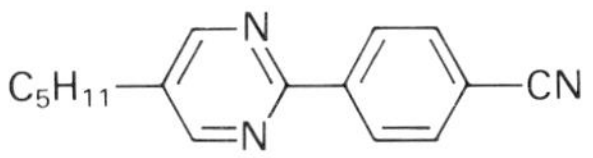

C–I 71 °C; N–I (52 °C)

15 [74]

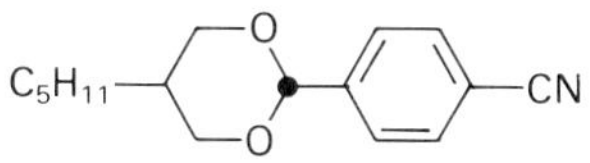

C–I 56 °C; N–I (52 °C)

16 [75]

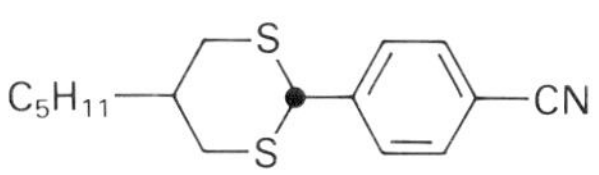

C–I 98 °C (supercools to 28 °C)

17 [76]

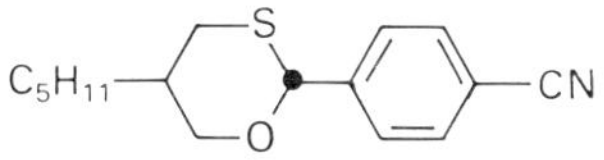

C–I 74 °C; N–I (19 °C)

18 [76]

phase, but both compounds **15** and **16** have a monotropic T_{N-I} at higher temperatures than for the corresponding biphenyl (**8**), and for the oxygen and sulphur heterocycles, lower clearing points are produced as the oxygen atom is replaced by the larger sulphur atom. Compounds **15** and **16** are useful materials for display devices because of their higher clearing points than for compound **8**; also compound **15** has a high $\Delta\varepsilon$ of +21.3 (at $T/T_c = 0.98$) and compound **16** has a relatively high $\Delta\varepsilon$ (about +11), a low Δn (= 0.08) and a moderate viscosity, and allows enantiotropic mixtures to be obtained because of its marked ability to depress melting points in mixtures. Both for compounds **15** and **16** the lone pairs of electrons on the nitrogen and oxygen atoms give additional dipoles which increase $\Delta\varepsilon$ relative to the values for K15 (compound **8**) and PCH5 (compound **11**) respectively (see page 32).

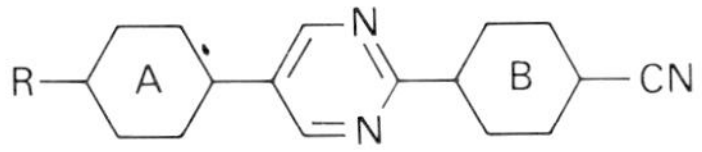

A = cyclohexyl, B = phenyl
A = phenyl, B = cyclohexyl
A = B = cyclohexyl

19

Three-ring terminal cyano compounds based on a central pyrimidine unit (compound **19**) have also been prepared and their transition temperatures, $\Delta\varepsilon$, viscosities and Δn have been measured[61]. A comparison of the transition temperatures for these series of compounds illustrates once again the effect of an alkyl chain substituted in an aryl or cyclohexyl ring and the effect of separating cyano and aryl units.

Numerous other examples of terminal cyano systems can be mentioned, and these include cyanoesters (see page 45), dimethylene linked compounds (see page

43), systems containing alkenyl[77,78], *trans*-1,2-cyclopropyl or *trans*-epoxy terminal groups[79], *cis-/trans*-1,3-cyclobutyl, 2,6-spiro(3,3)heptyl and 2,8-dispiro(3.1.3.1)decyl compounds[80] and compounds represented by structure **20**.

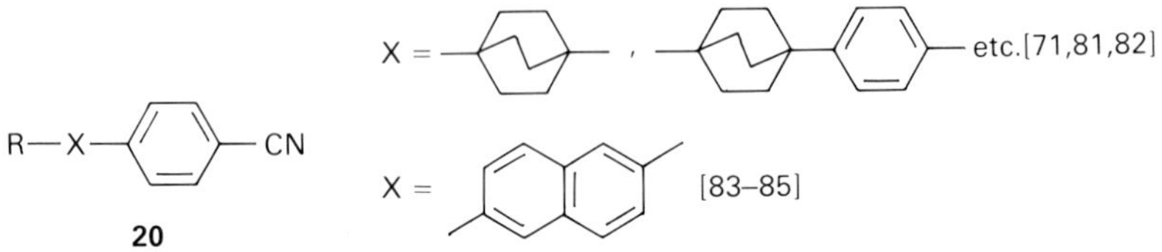

R—X—⟨benzene⟩—CN

20

X = bicyclooctane, bicyclooctane–phenylene — etc.[71,81,82]

X = naphthalene [83–85]

X = *trans*-decalin[86], tetralin[83,87]

2.4 Hydrocarbons

Some aspects of the properties of cyano-substituted mesogens were discussed earlier because of their central position in liquid crystal applications, but hydrocarbons should be the logical starting point for a discussion of the effect of structural change on mesomorphic properties, since hydrocarbons, uncomplicated by antiparallel correlations, dipole interactions, hydrogen bonding, etc., set the basis by which the influence of structural modifications can be judged.

The strength of the association of molecules is a balance of various forces which operate when molecules get close enough for their electron clouds to overlap; the dispersion forces are the main attractive forces in non-polar compounds[38]. The strength of the long-range interactions depends on the molecular separation, which is governed by steric factors, which therefore control the packing of the molecules and determine whether or not a layered structure is attainable. In the absence of lateral groups, a uniform type of molecular structure generally permits

C_3H_7–(cyclohexyl)–(cyclohexyl)–C_3H_7

C 64.2 °C (S 58 °C) S 81.8 °C I

21 [38]

C_3H_7–(cyclohexyl)–CH_2CH_2–(cyclohexyl)–C_3H_7

C 34.6 °C S 73.0 °C I

22 [88]

C_5H_{11}–(cyclohexyl)–CH_2CH_2–(cyclohexyl)–C_5H_{11}

C 46 °C S_B 109 °C I

23 [89]

the closest packing. For example, compound **21** is composed entirely of sp^3 hybridized carbon atoms and the regular zig-zag arrangement of chair cyclohexanes and alkyl chains permits packing in a layered structure giving S properties; commonly with cyclohexane systems, the S_B phase or other ordered S phase is produced. Compound **22**, with a dimethylene linking unit (see page 43), which would from one point of view be regarded as a serious disadvantage because of its lack of rigidity, also has a uniform structure giving good packing and is smectogenic. Increasing the alkyl chains in compound **22** to give compound **23** extends the zig-zag structure of the molecule and raises the clearing point, but still

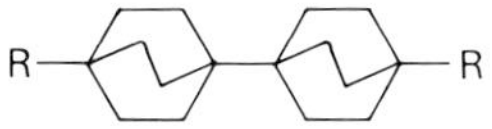

R—[bicyclo(2.2.2)octane]—[bicyclo(2.2.2)octane]—R

24 [90]

gives a smectic phase. For similar reasons, the compounds **24** with barrel-shaped bicyclo(2.2.2)octane rings and a collinear central unit would be expected to give good packing and indeed they show broad S ranges[90, 91]; the diethyl compound is believed to be the mesogen with the shortest molecular length. Even

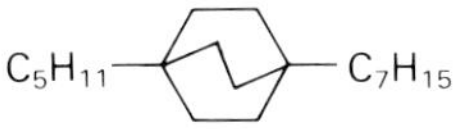

C_5H_{11}—[bicyclo(2.2.2)octane]—C_7H_{15}

25

one bicyclo(2.2.2)octane ring (see compound **25**) gives derivatives which have quite high virtual $T_{N\text{–}I}$ of about −30 °C[92]. All of these classes of compound show a rising curve for clearing point with initial increasing length of alkyl chain, suggesting that the chain is also important in molecular interactions and in these instances helps the packing. Beyond a certain length of alkyl chain, the advantage of better packing is more than offset by greater flexibility and the clearing temperatures decrease (see Table 2.3).

Table 2.3 Transition temperatures (°C) for 4,4′-di-n-alkylbibicyclo-(2.2.2)octanes (compound **24**)[90]

R	C–S	S–I
a...CH_3	196	—
b...C_2H_5	119	209
c...C_3H_7	155	222
d...C_4H_9	92	246
e...C_5H_{11}	46	247
f...C_6H_{13}	60	240
g...C_7H_{15}	47	226
h...C_8H_{17}	68	212

Biphenyls, terphenyls and related aromatic compounds, although not completely planar because of the repulsions between hydrogen atoms at *ortho* positions, have a flat, rigid core which favours a layered structure and gives S phases, as shown for compounds **26**, **27** and **28**. The series for compounds **27** and **28** have descending curves for plots of clearing points against length of alkyl

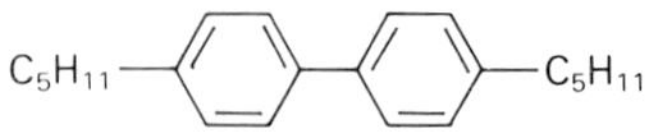

C 26 °C S 47.6 °C S 52.2 °C I

26 [38]

C 192 °C S 213 °C I

27 [93]

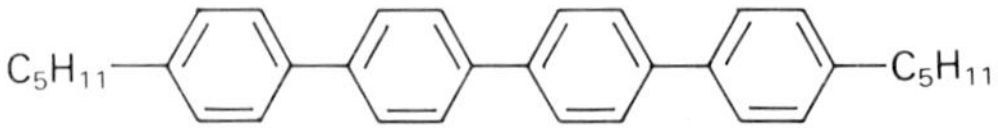

C 297 °C S_A 352 °C I

28 [94]

chain, and here the alkyl chains progressively destabilize the flat packing with which they are less compatible.

When aliphatic and aromatic systems are linked together in the core, they give molecules that cannot pack uniformly densely because of their intramolecular differences in shape, and the layer structure seems to be disturbed, producing N phases, with lower clearing points than those for the fully aromatic or aliphatic systems (compare compounds **29** with compounds in Table 2.4 and compound **29b** with compound **26**).

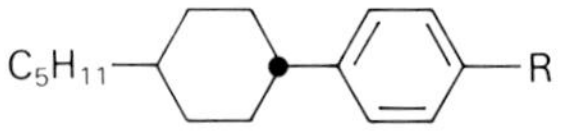

a...R = C_3H_7; C −12.7 °C N −11.2 °C I
b...R = C_5H_{11}; C −0.8 °C I; N–I (−5.0 °C), S–N (−8.0 °C)
c...R = C_7H_{15}; C 16 °C S 31 °C I

29 [38]

Table 2.4 Clearing temperatures (°C) for several dialkyl-biphenyls, -bicyclohexanes and -bibicyclo(2.2.2)octanes

R,R'	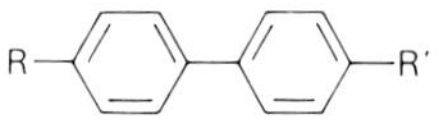 R–biphenyl–R'	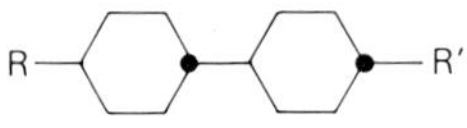 R–bicyclohexane–R'	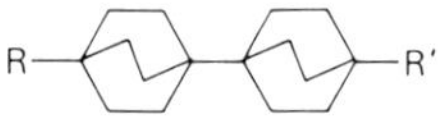 R–bibicyclo(2.2.2)octane–R'
a...C_3H_7, C_3H_7	61.2[a]	81.8[b]	222[e]
b...C_5H_{11}, C_5H_{11}	52.2[b]	110.4[a]	247[e]
c...C_7H_{15}, C_7H_{15}	—	—	226[e]
d...C_3H_7, C_7H_{15}	50.5[c]	97.8[d]	—

[a]Reference [95]. [c]Reference [96]. [e]Reference [90].
[b]Reference [38]. [d]Reference [97].

A comparison of the clearing temperatures (S–I) for specific sets of dialkyl-biphenyls, -bicyclohexanes and -bibicyclo(2.2.2)octanes is shown in Table 2.4 and these values clearly demonstrate that in molecules with structurally homogeneous cores the order of mesophase stability decreases in the order bicyclo(2.2.2)octane, cyclohexane, benzene. This order is one that has been established for many other systems[98], with only a few exceptions (some of which are discussed later), and given this order and the relationship between a high mesophase stability and structural uniformity, it is possible to rationalize the results shown in Figure 2.3. The T_{N-I} values for compounds **8**, **11** and **14** increase because a benzene ring is being replaced by a ring of superior mesophase stability, and the anti parallel correlation of compounds **11** and **14** (see Figure 2.4) mitigates the problem of packing different structural types; compounds **10** and **13**

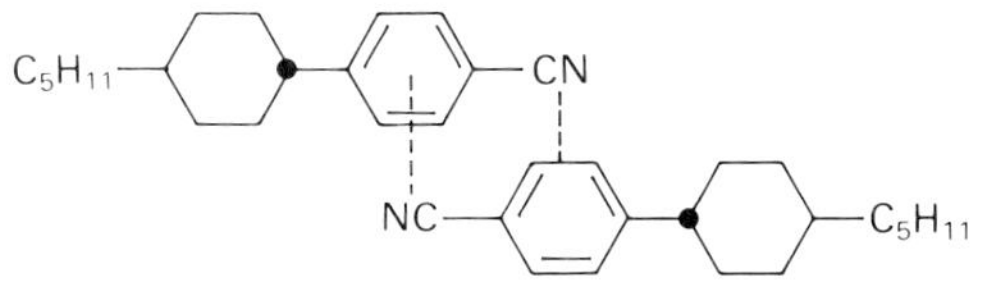

Figure 2.4 Antiparallel correlation in PCH5.

have even higher T_{N-I} values because the second phenyl ring is replaced. For compounds **9** and **12**, the structurally similar regions of alicyclic ring and alkyl chain are separated by a flat aromatic region with a resulting decrease in thermal stability. Bicyclo(2.2.2)octane is such an efficient unit in promoting mesophase stability, however, that when it replaces either phenyl ring (compounds **12** and **14**) it still gives compounds of higher clearing points than compound **8**. Eidenschink[99] has also discussed this situation and has proposed an alternative explanation for some of the results shown in Figure 2.3.

Table 2.5 Transition temperatures (°C) for some symmetrically linked cyclohexanes[89]

R	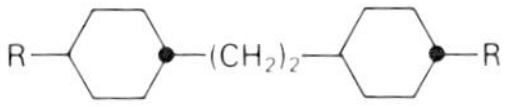 R–⟨⟩–$(CH_2)_2$–⟨⟩–R **30**	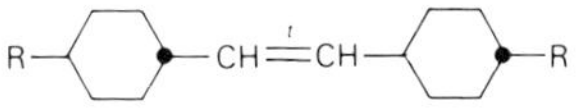 R–⟨⟩–CH=CH–⟨⟩–R **31**	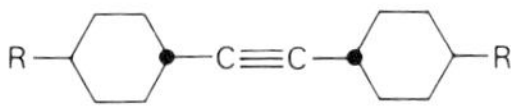 R–⟨⟩–C≡C–⟨⟩–R **32**
a...C_5H_{11}	C 46 S 109 I	C 53 S 95 I	C 52 I; N–I (50)
b...C_7H_{15}	C 52 S 115 I	C 39 S 106 I	C 55 S 57 I
c...C_9H_{19}	C 68.5 S 114 I	C 84.5 S 91 I	C 55.5 S 66 I

Compounds **32** (Table 2.5) were the precursors to compounds **31** and **30** and transition temperatures for these series show that the flat double bond has a smaller adverse effect than the linear triple bond, which in fact gives a greater separation of the two cyclohexane rings, but both unsaturated systems do show depressed thermal stabilities relative to the saturated compounds; compound **32a** shows a monotropic nematic phase.

2.5 Further comparison of benzene, cyclohexane, bicyclo(2.2.2)octane and cubane units

The values for the transition temperatures shown in Table 2.4 convincingly demonstrate the order of bicyclo(2.2.2)octane > cyclohexane > benzene in generating mesomorphic behaviour, and currently that order has been noted in a large number of systems of varied structures, including two- and three-ring compounds, esters, fluoro-substituted compounds, cyano- and non-cyano-substituted systems, etc. Gray[98] has shown some of the various systems for which this order applies and has noted some of the exceptions. Sometimes benzene is superior to both bicyclo(2.2.2)octane and cyclohexane, but this situation is confined to compounds where the presence of a benzene unit allows much greater

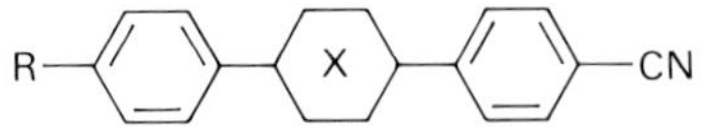

R—⟨benzene⟩—⟨X⟩—⟨benzene⟩—CN

33

RHN—⟨X⟩—⟨benzene⟩—CN

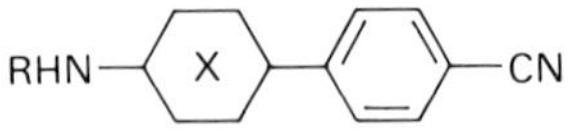

34

conjugation between parts of the molecule (e.g. as in compound **33**, X = benzene[43, 100], X = cyclohexyl[101] or where an alkoxy or alkylamino substituent can advantageously conjugate with an aromatic ring (e.g. as in compound **34**, X = benzene[46, 47], X = cyclohexyl[47]). Other exceptions

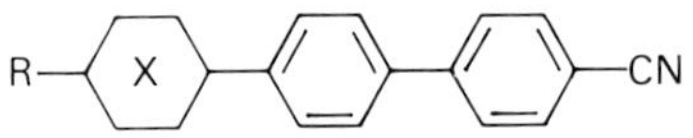

R—⟨X⟩—⟨benzene⟩—⟨benzene⟩—CN

a...X = benzene[43,45,100]
b...X = cyclohexane[101]
c...X = bicyclo(2.2.2)octane[71]

35

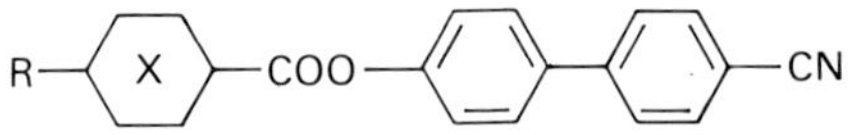

R—⟨X⟩—COO—⟨benzene⟩—⟨benzene⟩—CN

a...X = benzene[102,103]
b...X = cyclohexane[104,105]
c...X = bicyclo(2.2.2)octane[106]

36

to the general order are for three-ring systems of type **35** and **36** which give the order bicyclo(2.2.2)octane > benzene ⩾ cyclohexane (whereas the analogous two-ring systems give the normal order). This may arise because the fully aromatic molecules have extensive regions of delocalizable π electrons and the intrinsic advantage of a cyclohexane ring does not compensate for diminishing the extended, polarizable π-electron region; with bicyclo(2.2.2)octane, however, the balance is in favour of the alicyclic system.

An alternative explanation[104] which has been discussed in the context of other systems[39, 40, 47, 107]) suggests that because the three-ring systems have much higher T_{N-I} values than the two-ring systems (e.g. 220 °C for BICH5, i.e. compound **35**, X = cyclohexyl, R = C_5H_{11}; 55 °C for PCH5, i.e. compound **4**, R = C_5H_{11}), the former compounds containing cyclohexane rings may have significantly more of the *trans*-1,4-diaxial conformation present at the elevated temperatures close to the clearing point. Approximate calculations can be made to determine the magnitude of such a conformational change by using the conformational energy values for the substituents in the cyclohexane ring[108]. For compound **36b** (R = C_5H_{11}[104]) at its clearing point of 240 °C the proportion of diaxial component could be quite large (about 5 per cent), but for compound **35b** (R = C_5H_{11}[101]), at its clearing point of 220 °C, the diaxial component would be small (about 0.8 per cent); for the corresponding two-ring systems, 4-cyanophenyl *trans*-4-pentylcyclohexanecarboxylate (see Table 2.6, compound **43c(iii)**) and PCH5 (compound **4**, R = C_5H_{11}; see Table 2.2), the diaxial components are 1.4 and 0.1 per cent at the clearing points of 80 and 55 °C respectively. Although a greater proportion of material with a diaxial conformation will be detrimental to mesophase stability, less than 1 per cent would seem to be insufficient to account for the lower clearing point of compound **35b** (R = C_5H_{11}) relative to compound **35a** (R = C_5H_{11}), and other factors (see above, and possibly interannular rotation) must also be significant.

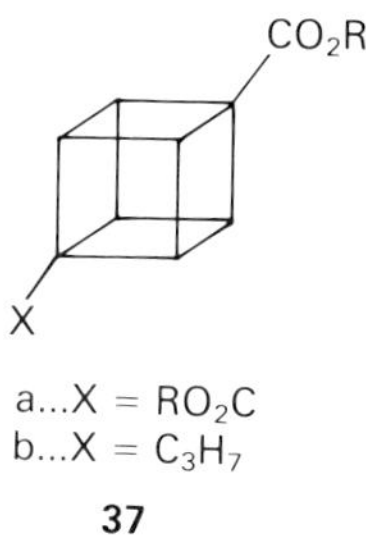

a...X = RO_2C
b...X = C_3H_7

37

Another system which, like benzene and bicyclo(2.2.2)octane, has collinear 1,4 bonds is cubane. However, cubane di- and mono-esters (**37a** and **37b**) have been shown to be markedly inferior to their bicyclo(2.2.2)octane, cyclohexane and benzene analogues in promoting mesophase thermal stability, and the clearing points of the cubane derivatives are more than 60 °C lower than those for the other three systems[109, 110]. Cubane is probably the worst of these systems, not because its molecular dimensions are significantly different from those of the other units, but because its rigid and angular structure prevents efficient packing; in support of this view, none of the cubane derivatives reported is smectogenic.

2.6 Dimethylene, oxymethylene and ester linking groups

Some of the earliest types of linking group to be used were —CH=N—, —N=N— and azoxy units, and studies of compounds with these linkages, particularly Schiff's bases, still continue[111–114]. Such linking units are favourable, from the point of view of mesophase thermal stability, because they are able to extend a region of conjugation, particularly when placed between two aromatic units, and because they confer a measure of rigidity to the molecule. They have the disadvantage, from the point of view of commercial exploitation in displays, in that they are sensitive to moisture and/or are photochemically unstable. Other π-type linking systems which have been examined include —C≡C—[115–117] and —CH=CH—[118, 119] and the readily prepared and useful ester link which is acceptable in most systems, including the following varied examples: 4-substituted-phenyl 4-substituted-benzoates[54, 120] and some naphthalene analogues[121], aryl-7-alkyl-9,10-dihydrophenanthrene-2-carboxylates[122], 4-substituted-phenyl[52, 54] and *trans*-4-alkylcyclohexyl[123] *trans*-4-alkylcyclohexane-1-carboxylates, 4-alkylphenyl, 4-alkoxyphenyl[82, 124–126] and cyanoaryl[82, 106] 4-alkylbicyclo(2.2.2)octane-1-carboxylates, and diesters[127, 128]; for numerous references to laterally substituted esters see page 50. Once the advantages of saturated alicyclic systems in generating compounds of high clearing points were realized, the compatibility of a dimethylene linking group within these systems was soon recognized[73, 104, 129], showing that even flexible links within a molecule may be acceptable. It is important, however, and directly contrary to the situation for the π systems referred to above, that the dimethylene link with its tetrahedral groups does not separate two flat regions of a molecule such as two aromatic rings. Some examples of molecules with dimethylene linkages are compounds **30** (Table 2.5) and compounds **69**, and further illustrations of the necessity of structural homogeneity are provided by comparing the $T_{N\text{-}I}$ values for compounds **38**, with n = 0 and 1, which show that as the ring system X becomes less saturated and more purely aromatic, so the dimethylene unit has an increasingly adverse effect[130].

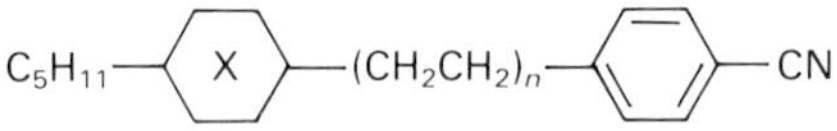

C_5H_{11}—(X)—$(CH_2CH_2)_n$—(phenyl)—CN

38

X	n = 0 $T_{N\text{-}I}$ (°C)	n = 1 $T_{N\text{-}I}$ (°C)
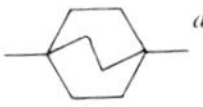 bicyclo[2.2.2]octane [a]	100	113
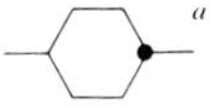 *trans*-cyclohexane [a]	55	51
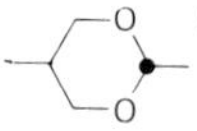 1,3-dioxane (O, O) [b]	(52)	(19)
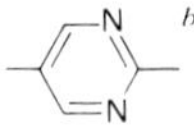 pyrimidine (N, N) [b]	(52)	[−11]
 benzene [a]	35	[−24]

[ ] Virtual transition.
( ) Monotropic transition.
[a] Reference [73].
[b] Reference [130].

A similar effect is also demonstrated when a dimethylene unit is inserted, not in the middle of a core, but between the core and the terminal cyano group, as in the

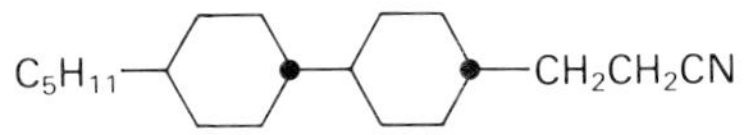

C_5H_{11}–cyclohexyl–cyclohexyl–CH_2CH_2CN

C 9 °C S 30 °C S 108.8 °C I

39 [107]

C_5H_{11}–cyclohexyl–cyclohexyl–CN

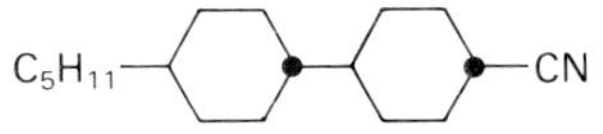

C 62 °C N 85 °C I; S–N (52 °C), S–S (43 °C)

40 [60]

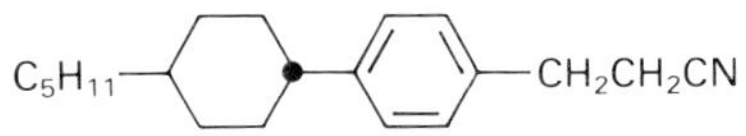

C_5H_{11}–cyclohexyl–phenyl–CH_2CH_2CN

C 44.8 °C I; S–I (28.3 °C)

41 [107]

C_5H_{11}–cyclohexyl–phenyl–CN

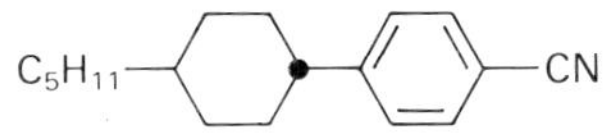

C 31 °C N 55 °C I

42 [59]

PCH and CCH systems[107]. For example, compound **39** shows an increased thermal stability with respect to CCH5 (**40**) when a dimethylene unit is located between the cyclohexyl ring and the cyano group, whereas the opposite effect is seen for compounds **41** and PCH5 (**42**).

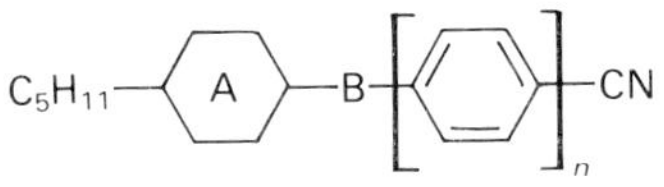

C_5H_{11}–(A)–B–[phenyl]$_n$–CN

43

Table 2.6 Transition temperatures (°C) for compounds of structure **43**

	A = benzene				A = cyclohexane				
	a…n = 1		b…n = 2		c…n = 1		d…n = 2		
B	C–I	N–I	C–N	N–I	C–N/I	N–I	C–S_A/N	S_A–N	N–I
(i) CH_2CH_2	62	[−24][a]	60	136[b]	30	51[c]	79	86	184[b]
(ii) CH_2O	49	[−20][a]	91	147[a]	74	(49)[d]	107	136	193[d]
(iii) COO	64	(55)[e,f]	109	237[e,f]	48	79[d,g]	89		240[d,h]

[ ] Virtual transition.
( ) Monotropic transition.
[a]Reference [132].
[b]Reference [133].
[c]Reference [73].
[d]Reference [104].
[e]Reference [102].
[f]Reference [103].
[g]Reference [52].
[h]Reference [105].

Cyano compounds (**43**) containing dimethylene, oxymethylene and ester links were compared by Gray and McDonnell[104] as part of a more general study which included a comparison of cyclohexyl and phenyl derivatives. Some of these results are shown in Table 2.6 and indicate the general superiority of cyclohexane to benene in promoting mesophase thermal stability, as discussed elsewhere (see page 41). They also show that the T_{N-I} values for the oxymethylene-linked compounds are markedly lower, particularly in the fully aromatic systems, than those for ester-linked compounds, whereas the T_{N-I} values for dimethylene and oxymethylene compounds are similar, and frequently the melting points for the dimethylene compounds are the lowest of all the linkages (also see reference [131] for a similar comparison of these linking groups in laterally fluoro-substituted benzonitriles).

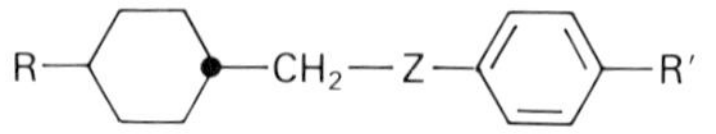

R–C₆H₁₀–CH_2–Z–C₆H₄–R′

a...Z = CH_2
b...Z = O

44

C_5H_{11}–bicyclooctane–X–C₆H₄–CH_3

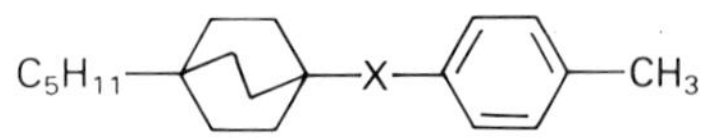

a...X = CH_2O; C–I 69 °C; N–I (33.5 °C)
b...X = OCH_2; C–I 63 °C; N–I [−210 °C]

45

Carr and Gray[132] have also studied the effect of the oxymethylene compared to the dimethylene link in systems of low polarity such as compound **44** and, in addition, give one remarkable example of the effect of reversing the central linkage, as shown in compound **45**. When the oxygen atom of the linkage is separated from conjugation with the aromatic ring so that separated flat or polar regions are created, there is a huge decrease in mesophase thermal stability. If, however, an oxygen atom is present as the only polar part of a molecule, then compounds of reasonable mesophase thermal stability may still

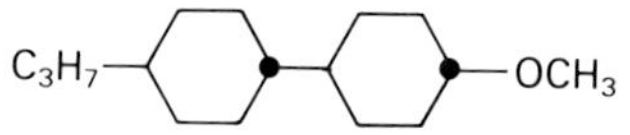

C_3H_7–C₆H₁₀–C₆H₁₀–OCH_3

C 10 °C N 17 °C I

46 [134,135]

C_3H_7–C₆H₁₀–CH_2O–C₆H₁₀–C_3H_7

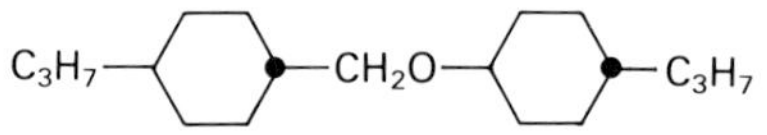

C 6.9 °C S 8.0 °C N 17.5 °C I

47 [136]

C_4H_9O–(tricyclic)–C_3H_7

C 53 °C N 110.6 °C I

48 [137]

C_3H_7–C₆H₁₀–C₆H₁₀–C_2H_5

C −5 °C S_B 67 °C I

49 [134]

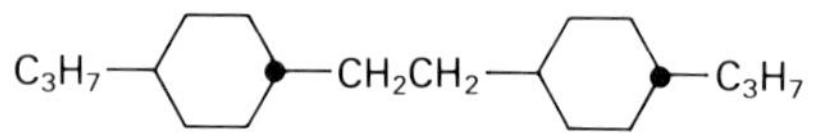

C_3H_7–C₆H₁₀–CH_2CH_2–C₆H₁₀–C_3H_7

C 34.6 °C S 73 °C I

50 [88]

C_5H_{11}–(tricyclic)–C_3H_7

C 76.8 °C N 136.5 °C I

51 [86]

be produced, as demonstrated by compounds **46**, **47** and **48**. When compared with their parent systems **49**, **50** and **51** respectively, it can be seen that although the clearing points have been depressed by up to 55 °C by the incorporation of an oxygen atom, the N thermal stability of compound **48** is still quite high and the totally S character of the parent systems **49** and **50** is eliminated; an enantiotropic N phase has been revealed in both cases (the virtual T_{N-I} value of compound **49** is, however, only a few degrees Celsius below the T_{S_B-I} value[134]). Compounds such as **46** are useful because of their low Δn (0.04) and low viscosity (for compound **46** the extrapolated viscosity from mixtures in ZLI

1132 is 7 cSt at 20 °C — a further indication of their much reduced S tendencies).

A comparison of the dimethylene, oxymethylene and ester links in completely saturated systems, where the possibility of mesomeric interactions does not arise, gives a rather different efficiency order from that shown in Table 2.6[38, 136]. The isolated oxygen atom of compound **47** lowers the efficiency of the packing of the parent system (**50**), giving an N phase of lower clearing point; the ester link (**52a**) behaves similarly, but the delocalized π system leads to somewhat greater interactions than in compound **47**, resulting in a higher clearing point.

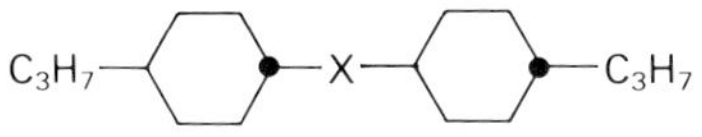

C_3H_7—⟨cyclohexane⟩—X—⟨cyclohexane⟩—C_3H_7

a...X = COO; C 22.8 °C N 36.6 °C I

52 [123]

2.7 Lateral substitution

2.7.1 Small lateral substituents

Investigations of the effect of lateral substituents on liquid crystalline behaviour may have several objectives, such as the desire to determine to what extent clearing points and the relative stabilities of different mesophases are affected by the sizes and types of lateral substituents or to find out how physical properties, e.g. dielectric properties, elastic constants, viscosity, etc., are modified; in the latter respect the fluoro and cyano substituents are particularly important. Recent reviews of this topic will be found in references [6, 7, 40].

Thermotropic liquid crystals are generally rod-like and even small lateral substituents usually perturb the structure and cause a significant depression in clearing point; therefore parent systems with high clearing points are required to allow the effect of a wide range of lateral substituents to be determined. Three-ring systems such as compounds **53**, **54** and **55** have recently been considered and the physical properties of **53** have been studied extensively [140, 141]. Both for compound **53**[141] (R = C_6H_{13}; X = CH_3, C_2H_5, C_3H_7, Cl, CN, CH_2CN, $COCH_3$) and for compound **55**[40] (X = CH_3, F, Cl, Br, CN, NO_2) a plot of clearing point against van der Waals volume of the substituent gives, with only minor deviations, a smooth curve to which even the strongly dipolar substituents conform and so suggests that dipolar attractive forces are much less important than dispersive forces in influencing N thermal stability; Gray *et al.*, however, were the first to demonstrate that a lateral dipolar substituent has a different effect on the thermal stability of S and N phases (see page 49). Fluorine, being the smallest substituent possible, has least effect in depressing the clearing point and a maximum depression of about 50 °C is found in non-cyano systems (see, for example, references[143, 144]). However, even a substituent as small as fluorine[144, 145] can have a much greater effect on S thermal stability by diminishing the efficiency of the packing of the molecules in

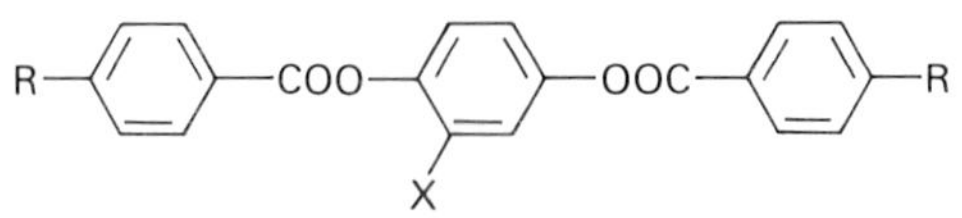

R—C6H4—COO—C6H3(X)—OOC—C6H4—R

53 [138–141]

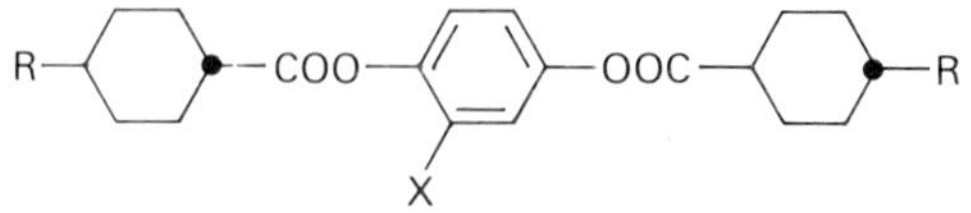

R—C6H10—COO—C6H3(X)—OOC—C6H10—R

54 [141,142]

C_5H_{11}—C6H10—C6H3(X)—C6H10—C_5H_{11}

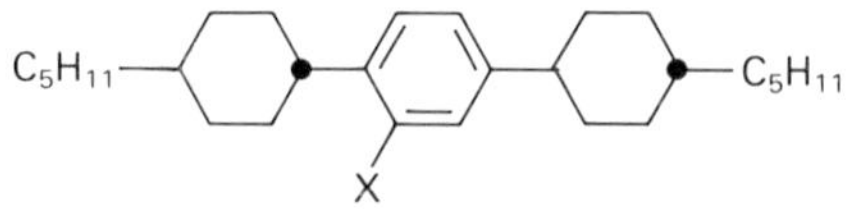

55 [40]

layers, and for system **55** the stability of the S phase is reduced by 117 °C. The ability of a lateral fluoro substituent to diminish the stability of S phases relative to N phases without significantly increasing viscosity is extremely useful in producing improved materials for electrooptic devices where the possible occurrence of S phases is a disadvantage (see page 55 and references [133, 144, 146]). Substituents larger than fluorine may destabilize the S phase to an even greater extent, but they would also give compounds with even lower clearing points and possibly with other adverse physical properties, such as increased viscosities.

The general variations outlined above do, however, depend on the nature of the parent system, and in some cases lateral substitution may increase molecular breadth and additionally cause increased twisting about an interannular bond[147], as in 2-substituted biphenyl or terphenyl stystems, or conversely the substitutent may be shielded by the molecular structure so that molecular breadth is not increased significantly. Two illustrations of the former situation are

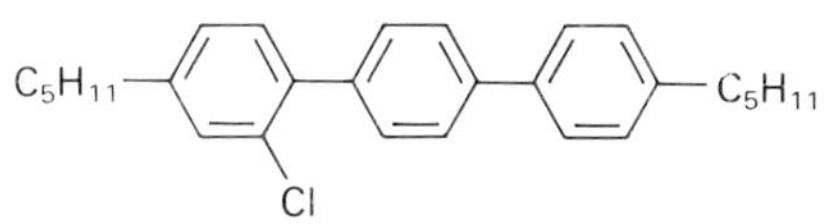

C_5H_{11}—C6H3(Cl)—C6H4—C6H4—C_5H_{11}

C 56 °C N 60 °C I
(Supercools to 25 °C as N phase)

56

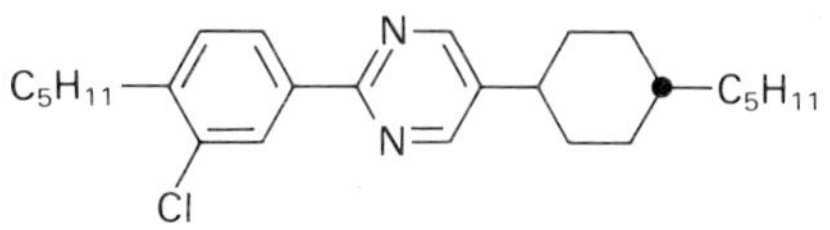

C_5H_{11}—C6H3(Cl)—pyrimidine—C6H10—C_5H_{11}

C 33 °C S 121 °C I

57

provided by compound **56**[40] which is purely nematogenic and shows that the clearing point of the parent system (C 192 °C S 213 °C I) has been depressed by 153 °C and the S thermal stability by at least 188 °C; comparison of compound of **55**[40] (X = Cl; C 46 °C N 96 °C I, no S phase down to −15 °C) with the parent compound (C 50 °C S 196 °C I) shows that in this instance S phase stability is decreased by at least 211 °C. These two examples contrast with other comparisons provided by Osman[40], such as compound **57**, in which the S phase of the parent system still persists and is only depressed by 68 °C (see also page 55 for the effect of four categories of fluoro substitution as described by Balkwill *et al.*[144]).

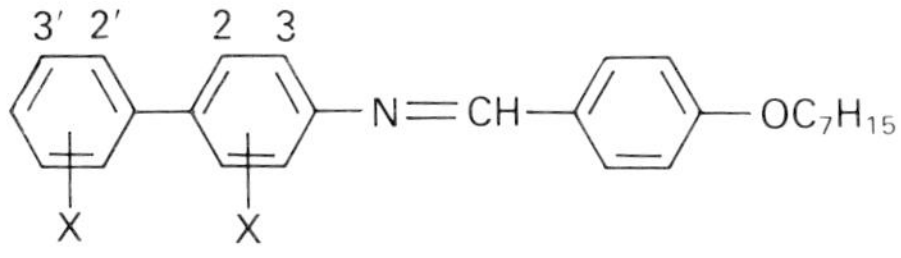

58 [143,148,149]

A further example of a twisting effect superimposed on a broadening effect is shown by the family of Schiff's bases **58** (*see* Table 2.7) for which the 2- and the

Table 2.7 T_{N-I} values for compounds **58**

X	T_{N-I} (°C)	ΔT(X–H) (°C)
H[150]	163.5	
2F	110	−53.5
2Cl	(45)	−118.5
2Br	(25)	−138.5
2I	(3.5)	−160.0
2′Cl	(52)	−111.5
2′Br	(36)	−127.5
2′I	(10.5)	−153
3Cl	(90)	−73.5
3Br	(82)	−81.5
3′Cl	136.5[a]	−27
3′Br	(124.5)[a]	−39

( )Monotropic transition.
[a] T_{S-I}.

2′- isomers give much larger depressions of T_{N-I}[143, 148, 149]. With the laterally substituted Schiff's bases **59**, the T_{N-I} values have been plotted against the sizes of a range of substituents (including X = NO_2) to give a curve showing that the transition temperatures decrease smoothly with increased substituent size (see page 47)[148]. The curve is steeper than that for the 3′-X-4′-n-octyloxy-biphenyl-4-carboxylic acids[151] in which the same substituents are used, but do not exert a twisting effect. This shows that the substituents act proportionately to their size in twisting and/or broadening a molecule. The curve for the T_{S-N} values of compound **59**, however, is much steeper than that for the T_{N-I} values and the point for X = NO_2 does not conform. The differences between these two plots are explained by bulky substituents having a more serious disturbing effect on the lamellar S order and by the counteracting influence of dipolar substituents

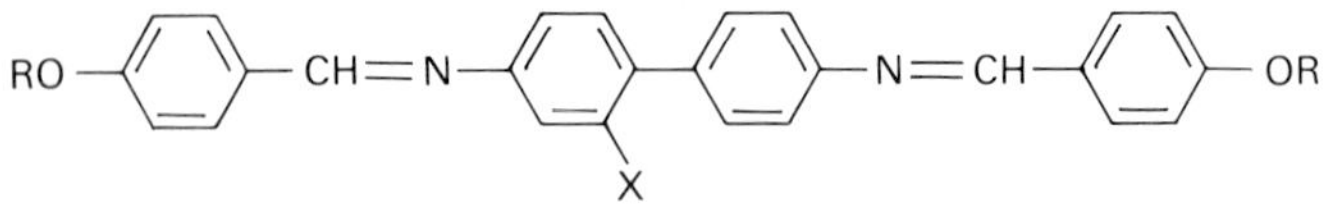

59

tending to increase S thermal stability; the different effect of a dipolar substituent on N–I and S–N transition temperatures has also been noted in 3′-X-4′-n-octyloxybiphenyl-4-carboxylic acids[151]. Di-, tri- and tetra-substituted derivatives of the parent system of compound **59** have been prepared and have shown that the effect of introducing more than one substituent into a mesogen is approximately additive[149].

On the other hand, instead of promoting a twisting/broadening effect, a lateral fluoro substituent may be shielded by the structure of the core as with

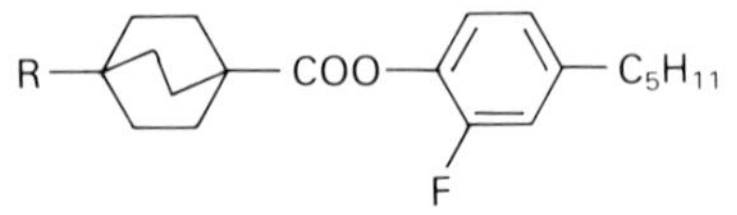

60

compounds **60** and the effect of 2-fluoro substitution on $T_{N\text{–}I}$ is then remarkably small (average decrease about 3 °C) and in some cases $T_{N\text{–}I}$ actually increases with fluorine substitution[152]. The $T_{N\text{–}I}$ values for 2-fluoro-4-cyanophenyl-4-alkylbicyclo(2.2.2)octane-1-carboxylates are consistently higher than those for the parent compounds by an average of 25 °C[153] and frequently *trans*-4-alkylcyclohexanecarboxylates show the same effect[154]; in addition to a shielding effect in these cases it is likely that the association of the terminally cyano-substituted molecules is affected by fluorine substitution to give a modified type and extent of molecular association (antiparallel correlation) which leads to a greater length-to-breadth ratio and hence to higher clearing points.

The clearing temperatures for the unsubstituted, 2-fluoro- and 3-fluoro-4-pentylphenyl esters derived from bicyclo(2.2.2)octane-, cyclohexane- and

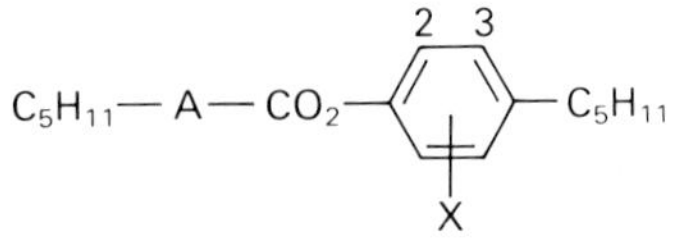

61

benzene-carboxylic acids are shown in Table 2.8 (compounds **61**); 2-fluoro substitution has a different effect in each system because of greater shielding in the order benzene, cyclohexane, bicyclo(2.2.2)octane, whereas 3-fluoro substitution has a similar effect in all three systems. Even a 2-chloro or a 2-cyano substituent is substantially shielded in the bicyclo(2.2.2)octane esters and small depressions in $T_{N\text{–}I}$ of only about 27 and 31 °C respectively

Table 2.8 Comparison of clearing points for compounds **61**

A	X	T_{N-I} (°C)	ΔT_{N-I}(2F—H) (°C)	ΔT_{N-I}(3F—H) (°C)
 [bicyclooctane ring]	H[a]	64.5		
	2F[b]	65	+0.5	
	3F[b]	(38.5)		−26.0
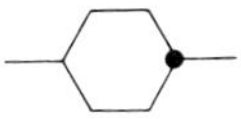 [cyclohexane ring]	H[c]	48		
	2F[d]	36.5	−11.5	
	3F[d]	26.3		−21.7
 [benzene ring]	H[e]	(25.9)		
	2F[d]	(1.2)	−24.7	
	3F[d]	[0]		−25.9

( ) Monotropic transition.
[ ] Virtual transition.
[a]Reference [125].
[b]Reference [152].
[c]Reference [54].
[d]Reference [155].
[e]Reference [156].

occur[152], whereas these substituents give depressions in clearing points of 44 and 55 °C respectively in compound **53** and of 100 and 116 °C respectively in compound **55**.

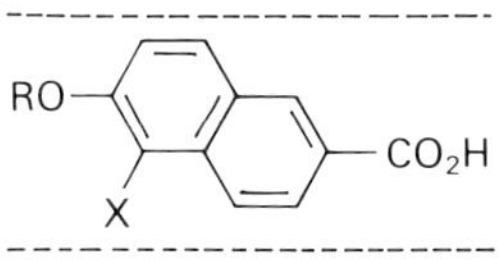

RO— [naphthalene] —CO_2H, X

62

The 6-n-alkoxy-5-halogeno-2-naphthoic acids (**62**) provided the first classic illustration of how the halogeno substituent is shielded by the molecular structure and in the case of chloro and bromo gives increased mesophase stability of 11.7 and 8.5 °C respectively; even the large iodo substituent leads to an average clearing point depression of only 1 °C[157].

An important potential use for laterally substituted compounds lies in producing nematogens suitable for high information content, highly multiplexed twisted nematic displays. For this use a steep electrooptic contrast curve is required and the materials used should have low values of $\Delta\varepsilon/\varepsilon_\perp$ (<1) and of k_{33}/k_{11} (<1)[158–162]. The values of both of these parameters are too high with terminal cyano (polar) compounds and so non-polar (terminal dialkyl or alkyl-alkoxy) compounds, which have smaller $\Delta\varepsilon/\varepsilon_\perp$ and k_{33}/k_{11} ratios, are commonly added to give acceptable mixtures for matrix-addressed twisted nematic displays[159, 161]. However, an undesirable consequence of mixing polar and non-polar compounds is that the phenomenon of injected S phases may arise, i.e. although neither component shows an enantiotropic or monotropic S phase, certain compositions of the binary mixture may reveal an S phase (usually S_A, and

C_5H_{11}–(cyclohexyl)–(phenyl)–COO–(phenyl, X, Y)–CN

63

	T_{red}	$\varepsilon_{\parallel}$	$\varepsilon_{\perp}$	$\Delta\varepsilon$	$\Delta\varepsilon/\varepsilon_{\perp}$
a...X = H, Y = H; C 111 °C N 225 °C I	0.77	19.0	4.7	14.3	3.04
b...X = H, Y = CN; C 85.2 °C N 143.9 °C I	0.86	35.7	7.5	28.2	3.76
c...X = CN, Y = H; C 132.2 °C N 178.7 °C I	0.89	11.0	7.8	3.2	0.41

less commonly S_B or S_E)[163–165]. Lateral substitution of cyano or fluoro groups may simultaneously reduce $\Delta\varepsilon/\varepsilon_{\perp}$ (by increasing the dielectric constant perpendicular to the molecular long axis), reduce k_{33}/k_{11} (by increasing the width to length ratio) and suppress any S tendencies (by destabilizing the layer structures). The example in compound **63**[159] shows how the lateral cyano group can lead to a compound with a very low $\Delta\varepsilon/\varepsilon_{\perp}$ value, but with a much reduced $\Delta\varepsilon$ value; however, mixtures of **63b** and **63c** should give an acceptable value of $\Delta\varepsilon$ without the risk of injected S phases. More commonly, the effect of a lateral fluoro substituent has been considered for the same purpose and also as a way of affecting the antiparallel correlation of molecules. With aromatic terminal cyano systems, such as Ks, PCHs and terminal cyano-esters, their antiparallel correlations lead to a reduced $\Delta\varepsilon$ and an increased clearing point relative to that expected for the hypothetical unassociated molecules. However, a fluorine atom *ortho* to the cyano group, e.g. in 4-cyanophenyl benzoates **64** (see Table 2.9), decreases the molecular association and so gives a lower clearing point but an increase in $\Delta\varepsilon$ because of the reduced antiparallel correlation and the increased dipole; a fluoro substituent *meta* to the cyano group has a less significant effect.

The most remarkable example of the ability of an *ortho*-fluoro substituent to generate compounds of high $+\Delta\varepsilon$ arises in 4-cyano-3-fluorophenyl 4-alkylbenzo-

Table 2.9 Transition temperatures (°C) for some fluorinated 4-cyanophenyl 4-alkylbenzoates **64**

	R = C_5H_{11}		R = C_7H_{15}	
	C–N/I	N–I	C–N	N–I
a...X = H, Y = H	63.4	(56.8)[a,b,c]	44	56.5[b,c]
b...X = F, Y = H	53.5	55[a,c,d]	47	54[c,d]
c...X = H, Y = F	30	(20)[a,d,e]	28	28.5[d,e,f]

[a]Reference [40]. [d]Reference [166].
[b]Reference [103]. [e]Reference [167].
[c]Reference [153]. [f]Reference [162].

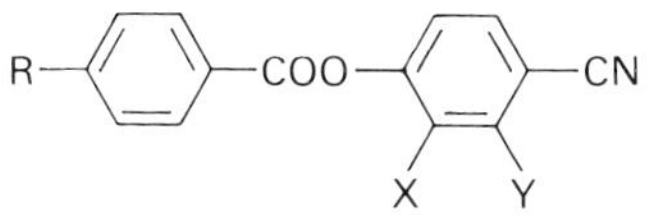

R—⟨benzene⟩—COO—⟨benzene(X, Y)⟩—CN

64

ates[162, 166]. For compound **64c** (R = C_7H_{15}), the 3-fluoro substituent has depressed T_{N-I} by 28 °C but has given a compound with an extremely high $\Delta\varepsilon$ of +48.9 (at T_{red} = 0.95), the largest value yet reported for an N material, and the dipole moment of the molecule appears to be entirely uncompensated by antiparallel correlation. An alternative interpretation of these results and a different view of the association between polar molecules has been advanced which proposes that there is an equilibrium between monomers and dimers in polar mesogens, and the dimers may have antiparallel or parallel orientation of their dipoles[168]. Esters such as **64c** may be highly associated, but the dimers may have only a small preference for antiparallel over parallel local order because the molecular shape and the distribution of other dipole centres prevents significant antiparallel dipole–dipole correlations of the cyano group.

The rotational viscosity of compound **64c** (R = C_7H_{15}) offsets the advantages provided by its high $\Delta\varepsilon$ so that it is unsuitable as a major component in twisted nematic displays; however, it could still be useful as an additive in low concentrations to lower the threshold voltage of N mixtures[162].

The effect of fluorine substitution has also been considered in K, M[133] and PCH[159] systems. For the first two systems even more positions for fluoro substitution of aromatic rings are possible and quite distinct effects of a 2- and 2′-fluoro substituent on T_{N-I} have been noted[133] and the 4-n-alkyl-4′-cyano-2-fluorobiphenyls give compounds of higher $\Delta\varepsilon$ than the parent system[169].

R—⟨cyclohexane⟩—⟨cyclohexane⟩(CN)(R′)

R = R′ = C_5H_{11}; C 25 °C S_B 30 °C N 66 °C I

65

The ability of the molecular core to shield a lateral substituent in a different way has been applied very effectively in compounds **65**, where the linear cyano group with its small spatial demand readily becomes the axial substituent in preference to the alkyl group, and is then shielded to a considerable degree by the other axial hydrogen atoms of the cyclohexane ring[134, 170]. These compounds have $\Delta\varepsilon = -8$ to -10 and a low Δn (~0.03), and they should be of use in positive contrast guest–host displays. The use of a lateral axial cyano group is capable of even further elaboration and two- and three-ring compounds, e.g. compound **66**, have been reported[134]; the transition temperatures for the parent compound related to compound **66** are not

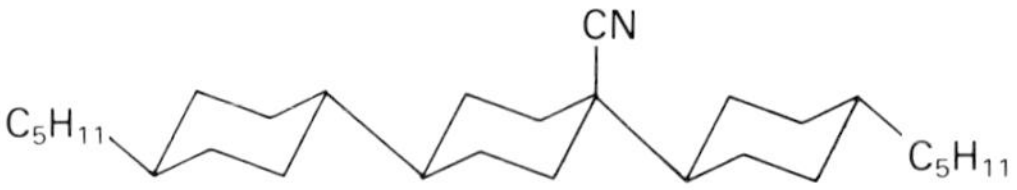

CN

C_5H_{11} — [tercyclohexane] — C_5H_{11}

C 33 °C S_B 176 °C S_A 185 °C N 199 °C I

66

available, but the closely similar 4,4″-di-heptyltercyclohexane has C 74 °C S 245 °C I[171].

Currently most liquid crystal devices are based on the twisted-nematic principle and require material of high positive dielectric anisotropy, but other devices such as guest–host displays with positive contrast and ECB (electrically controlled birefringence) devices require nematics of negative dielectric anisotropy. Axial cyano compounds **65** and **66** are such compounds, but alternative systems with lateral disubstituents have also been considered and the results for mono- and dicyano-substituted esters for all four combinations of phenyl or *trans*-cyclohexyl

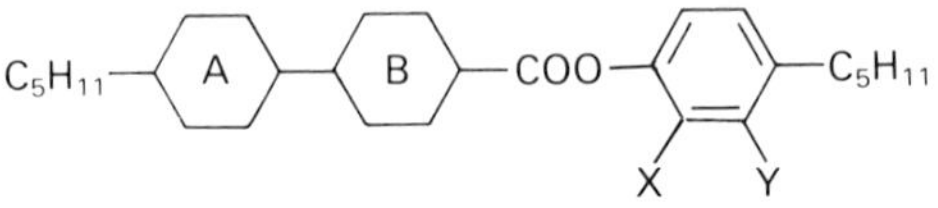

C_5H_{11}—(A)—(B)—COO—[benzene(X, Y)]—C_5H_{11}

a[172]...X = H, Y = H; C 95 °C S 152 °C N 176 °C I
b[173]...X = CN, Y = H; C 85 °C N 105 °C I
c[174]...X = CN, Y = CN; C 99 °C N 112.5 °C I

67 (A,B = benzene) [40]

as rings A and B in compound **67** have been reported[40]. The results differ slightly for each of these combinations of ring systems but, for example, for the fully aromatic compound the clearing point is decreased by monosubstitution (**67b**) and the S phase is suppressed, but disubstitution leads to an increased clearing point. The general conclusion is that the second of two identical substituents has little effect on the clearing point because it does not increase molecular separation and is shielded by the first substituent (compare compounds **60** and **62**), but it can influence the type of mesophase formed because of changes in the intermolecular forces. The lateral dicyano esters have large $\Delta\varepsilon$ of about −12 (e.g. −11.7 at T_{red} = 0.95 for the fully aromatic compound **67c**[174] and −13 for the analogous ester of the cyclohexylbenzoic acid at T_{red} = 0.92[175]), and although they are chemically and photochemically stable, they are of high viscosity. In an attempt to obviate this problem, 3,6-disubstituted

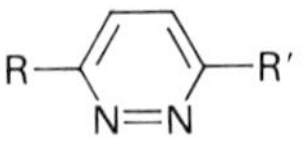

R—[pyridazine N=N]—R′

68

pyridazines (**68**) have been considered in the hope that the lone pairs of electrons on the nitrogen atoms would provide the required dipole orthogonal to the molecular long axis without increasing the molecular breadth[176–178]. These pyridazine derivatives are compounds of negative $\Delta\varepsilon$ (~ -9.3)[177], but their long-term photostability is unsatisfactory[178].

Not only is fluoro substitution a useful way of reducing antiparallel correlation in terminal cyano systems to give material with high $+\Delta\varepsilon$ values and reduced S tendencies (see page 52), but also fluoro substitution in non-polar hydrocarbons and in esters has produced materials with a variety of valuable physical properties; the effect of fluorination of N systems[144] and non-polar biphenyls and phenylpyrimidines[160] has been discussed. The first paper categorizes and discusses the incorporation of fluorine as a terminal group, in the terminal chain, or as a lateral substituent either close to the centre or pointing from the end of the molecular core.

When the lateral fluorine points towards the centre of the molecular core, then S phases are eliminated or strongly depressed and T_{N-I} is only slightly affected, whereas lateral substitution at the end of the core produces no significant reduction in, or may even enhance, S phase stability and T_{N-I} shows a greater depression than for the central substituent. These general guidelines have been successfully applied in producing several series of fluorinated N hydrocarbons of

C_3H_7–(cyclohexane)–CH_2CH_2–(benzene, X)–(benzene, Y)–C_3H_7

	a...X = H, Y = H;	C 67 °C S 119 °C N 144 °C I
I′33:	b...X = F, Y = H;	C 59 °C N 108 °C I; S (34 °C) N
I33:	c...X = H, Y = F;	C 40 °C N 108 °C I; no S phase detected on cooling

69

low viscosity. Examples of the I and I′ compounds are shown in **69** (**c** and **b** respectively); in each series wide N ranges are produced and S phases are almost totally absent, although the I′ series (see **69b**) does show greater S tendencies. The usefulness of the I compounds in display applications is demonstrated by the formation of mixtures (doped to be $\Delta\varepsilon$ positive with cyanobiphenyls) of broad nematic range (e.g. −50 to +111 °C) which at 5 V may have a response time, even at −20 °C, of about 500 ms[144].

Other examples of fluoro substitution in hydrocarbons are provided by compounds **70** (the difluoro compound **70c** shows a further depression of clearing point and a rather large increase in viscosity)[134, 144], compound **71**[144], compounds **72**[134], compounds **73** (also the dialkoxy and alkyl/alkoxy compounds)[146] and compounds **74**[180]. An interesting point to emerge from the results for the series of compounds represented by **73** and **74** is that the compounds with the fluoro substituent pointing towards the smaller terminal alkyl group (e.g. **73c** and **74b**) appear to cause a smaller depression of S thermal stabilities, whereas either type of fluorine substitution has approximately the

a...X = H, Y = H; C 34 °C S 146 °C N 164 °C I; η_{20} (cSt) = 20

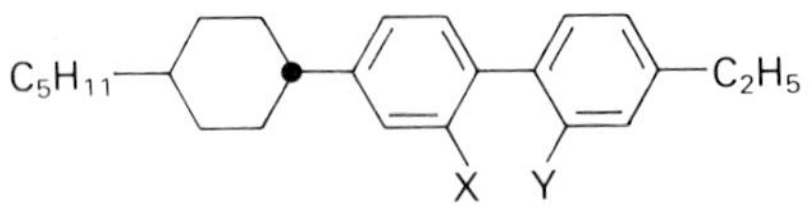

b...X = F, Y = H; {C 26 °C N 107 °C I[144,179]; / C 37 °C N 117 °C I[134];} η_{20} (cSt) = 24

c...X = F, Y = F; C 27 °C N 78 °C I[179]; η_{20} (cSt) = 42

70

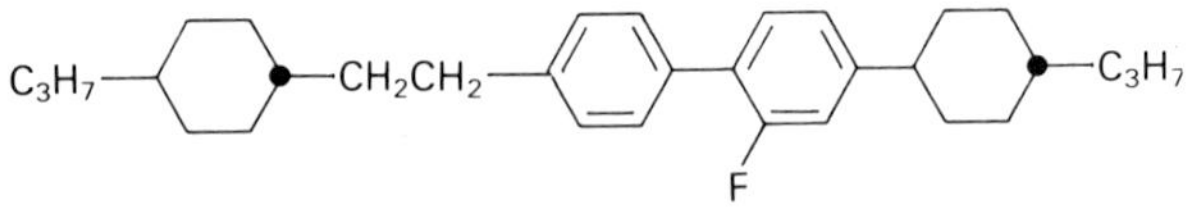

C 107 °C N 244 °C I

71

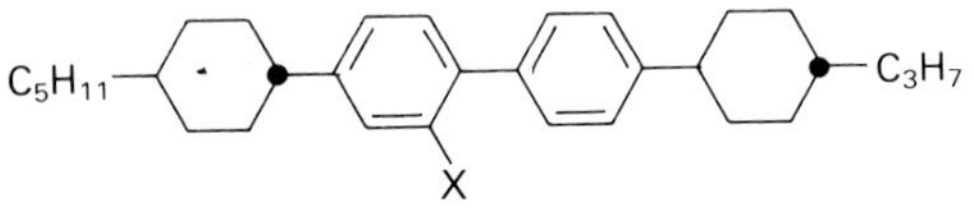

a...X = H[179]; C 58 °C S_B 232 °C S_A 251 °C N 311 °C I
b...X = F[134]; S_H 68 °C S_B 154 °C N 283 °C I

72

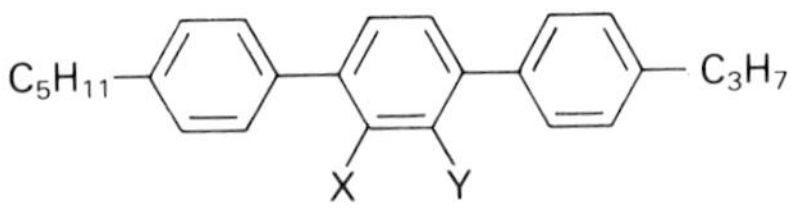

a...X = H, Y = H; C 180 °C S_E 200 °C S_B 214 °C S_A 218 °C I
b...X = F, Y = H; C 50 °C N 141 °C I
c...X = H, Y = F; C 61 °C S_A 99.5 °C N 141.5 °C I

73

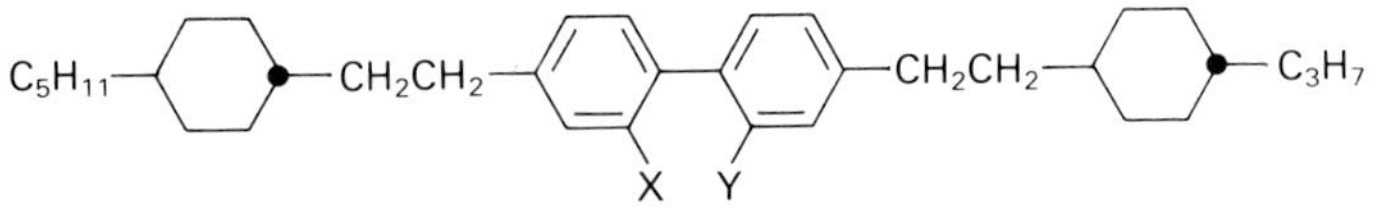

a...X = H, Y = H; C 66 °C S_J 147 °C S_B 191 °C S_A 214 °C N 227 °C I
b...X = F, Y = H; C 69 °C S_J 81 °C S_B 110 °C S_A 139 °C N 206 °C I
c...X = H, Y = F; C 76 °C S_B 103.5 °C S_A 123 °C N 207 °C I

74

same effect on the clearing points; a similar situation is probably shown by the series represented by compounds **69**.

Lateral substitution normally refers to substitution in the core of the molecule but, finally, it is convenient at this point to refer briefly to substitution in a terminal chain to give branched terminal groups. Gray and Kelly[181] have compared 4-alkyl- and 4-alkoxy-4′-cyanobiphenyls and 4-alkyl-4″-cyanoterphenyls in which a methyl group is systematically moved along the carbon chain; their results are similar to those in earlier work with different types of mesogen[182]. The depression in T_{N-I} gradually decreases as the methyl substituent is moved away from the molecular core and is attached to an increasingly flexible part of the chain, which therefore allows the substituent to be more readily accommodated in the normal rotational volume of the alkyl chain. Such chain branching will, in certain instances, generate chiral atoms and the corresponding enantiomers could give cholesteric systems for which the helical twist sense could be predicted by using the rules devised by Gray and McDonnell[183].

2.7.2 *Large lateral substituents*

The work discussed above has involved non-flexible substituents (e.g. F, CN) although some groups (e.g. C_2H_5, C_3H_7) with possible conformational variations were included as substituents in compound **53**[141]. Some work has also been carried out recently on compounds with very much larger lateral substituents[138, 139]. Using the same parent system as in **53**, a variety of

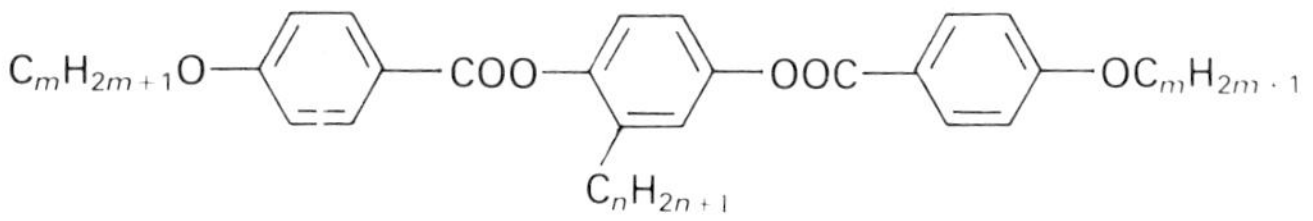

$C_mH_{2m+1}O$—C₆H₄—COO—C₆H₃(C_nH_{2n+1})—OOC—C₆H₄—OC_mH_{2m+1}

75

m	n	C–N (°C)	N–I (°C)
6	0	126	211
6	1	88	172
6	2	60	132
6	6	61	83
6	7	61	80
6	16	61	65

diesters **75** with combinations of terminal groups (mainly alkoxy [m = 1–10] and alkyl) and with lateral alkyl groups (n = 0–12 and 16) have been prepared. For compounds **75** with m = 6, for example, a lateral methyl group decreases the clearing point by 39 °C and another carbon atom (ethyl group) gives a further depression of 40 °C, but the difference between the clearing points for the hexyl and heptyl lateral groups is only 3 °C, and the clearing points are steadily converging to a limiting value for the clearing point of about 65 °C, as found for compound **75** (n = 16), which still shows an enantiotropic phase. All of the compounds studied are N in phase behaviour, and these long lateral chains seem very effectively to suppress S properties. Presumably the early members of the homologous series of lateral groups have the major effect in separating the aromatic cores, and with longer lateral chains the separation does not increase markedly because the lateral group can adopt a conformation along the molecular axis rather than across it[138, 139]. An interesting difference between lateral substitution and terminal substitution is that with a constant lateral group there is pronounced alternation in clearing points for an homologous series of terminal substituents (see page 30), but with a constant terminal group the homologous series of a lateral substituent does not produce an alternation. Similar effects have been observed recently for oxime benzoates (e.g. compound **76**) with long chain lateral alkyl groups[184].

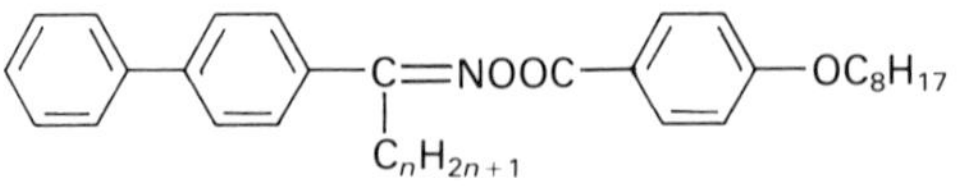

C=NOOC — OC_8H_{17}

C_nH_{2n+1}

76

2.8 Miscellaneous systems

New classes of mesogenic compounds continue to be revealed and recent examples of these include optically active compounds with axial chirality (**77**) (giving chiral S or chiral N phases, in some cases at room temperature), heterocyclic boron-containing compounds (**78**), and other more conventional heterocyclic systems such as compounds **79** and **80**; a review of some heterocyclic analogues of biphenyl mesogens has also appeared[189].

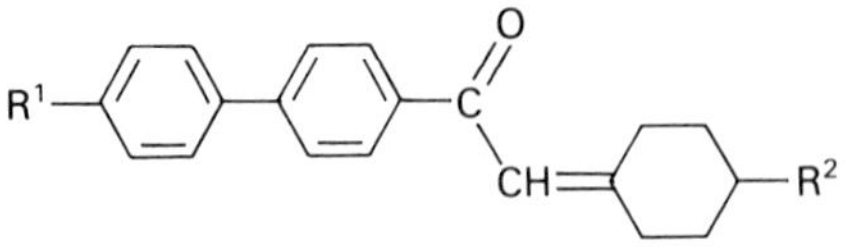

R^1 — C(=O) — CH= — R^2

R^1 = R
R^2 = R, OR, OCOArX, CO_2R

77 [185]

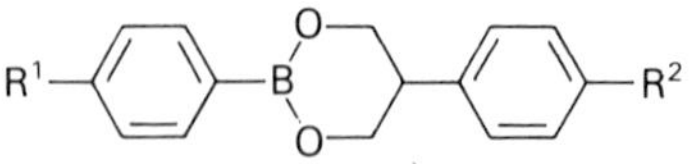

R^1 — B(O)(O) — R^2

R^1 = CO_2H, CN, CO_2Me, OMe
R^2 = R, OR

78 [186]

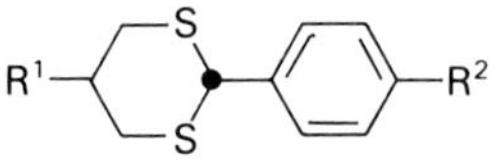

R^1 — S S — R^2

R^1 = R
R^2 = R, OR, CN

79 [187]

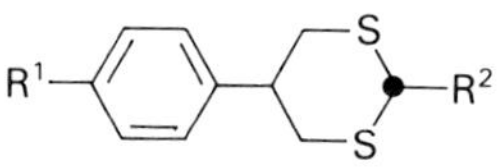

R^1 — S S — R^2

R^1 = OR, R
R^2 = R, ArX

80 [188]

2.9 References

1. F. Reinitzer, *Monatsh. Chem.*, **9**, 421 (1888).
2. D. Demus, H. Demus and H. Zaschke, *Flüssige Kristalle in Tabellen*, VEB Deutscher Verlag für Grundstoffindustrie, Leipzig, German Democratic Republic (1974).

3. D. Demus and H. Zaschke, *Flüssige Kristalle in Tabellen*, Vol. II, VEB Deutscher Verlag für Grundstoffindustrie, Leipzig, German Democratic Republic (1984).
4. W.E. Haas, 'Liquid crystal display research: The first fifteen years', *Mol. Cryst. Liq. Cryst.*, **94**, 1 (1983).
5. B. Bahadur, 'Liquid crystal displays', *Mol. Cryst. Liq. Cryst.*, **109**, 1 (1984).
6. G.W. Gray, 'Molecular geometry and the properties of non-amphiphilic liquid crystals', *Adv. Liq. Cryst.*, **2**, 1 (1976).
7. G.W. Gray, 'The chemistry of liquid crystals', *Philos. Trans. R. Soc. London, A*, **309**, 77 (1983).
8. G.W. Gray and P.A. Winsor (eds.), *Liquid Crystals and Plastic Crystals*, Vol. 1, Ellis Horwood, Chichester, England (1974).
9. G.W. Gray and P.A. Winsor (eds.), *Liquid Crystals and Plastic Crystals*, Vol. 2, Ellis Horwood, Chichester, England (1974).
10. G.W. Gray, *Molecular Structure and the Properties of Liquid Crystals*, Academic Press, New York (1962).
11. E.B. Priestley, P.J. Wojtowicz and Ping Sheng (eds.), *Introduction to Liquid Crystals*, Plenum Press, New York (1974).
12. S. Chandrasekhar, *Liquid Crystals*, Cambridge University Press, Cambridge (1977).
13. S. Chandrasekhar, B.K. Sadashiva and K.A. Suresh, *Pramana*, **9**, 471 (1977).
14. S. Chandrasekhar, 'Liquid crystals of disclike molecules', *Adv. Liq. Cryst.*, **5**, 47 (1982).
15. C. Destrade, N.H. Tinh, H. Gasparoux, J. Malthete and A.M. Levelut, *Mol. Cryst. Liq. Cryst.*, **71**, 111 (1981).
16. C. Destrade, P. Foucher, H. Gasparoux, N.H. Tinh, A.M. Levelut and J. Malthete, *Mol. Cryst. Liq. Cryst.*, **106**, 121 (1984).
17. B. Kohne and K. Praefcke, *Angew. Chem., Int. Ed. Engl.*, **23**, 82 (1984).
18. K. Ohta, A. Ishii, I. Yamamoto and K. Matsuzaki, *J. Chem. Soc., Chem. Commun.*, **1984**, 1099.
19. C. Piechocki and J. Simon, *J. Chem. Soc., Chem. Commun.*, **1985**, 259.
20. G.W. Gray and J.W. Goodby, *Smectic Liquid Crystals*, Leonard Hill, Glasgow, Scotland (1984).
21. D. Demus, S. Diele, S. Grande and H. Sackmann, 'Polymorphism in thermotropic liquid crystals', *Adv. Liq. Cryst.*, **6**, 1 (1983).
22. G.W. Gray in *The Molecular Physics of Liquid Crystals*, ed. by G.W. Gray and G.R. Luckhurst, Academic Press, London (1979), Ch. 12, p. 263.
23. H. Kelker, R. Hatz and C. Schumann, *Handbook of Liquid Crystals*, Verlag Chemie, Weinheim, Deerfield, Ill. (1980).
24. H. Schubert and H. Dehne, *Z. Chem.*, **12**, 241 (1972).
25. H.-J. Deutscher, H.-M. Vorbrodt and H. Zaschke, *Z. Chem.*, **21**, 9 (1981).
26. W. Maier and A. Saupe, *Z. Naturforsch.*, **13A**, 564 (1958).
27. W. Maier and A. Saupe, *Z. Naturforsch.*, **14A**, 882 (1959).
28. W. Maier and A. Saupe, *Z. Naturforsch.*, **15A**, 287 (1960).
29. W.H. de Jeu, J. van der Veen and W.J.A. Goossens, *Solid State Commun.*, **12**, 405 (1973).
30. E.T. Samulski, *Isr. J. Chem.*, **23**, 329 (1983).
31. A.G. Avent, J.W. Emsley, S. Ng and S.M. Venables, *J. Chem. Soc., Perkin Trans.*, *2*, **1984**, 1855.
32. G.R. Luckhurst in *Nuclear Magnetic Resonance of Liquid Crystals*, ed. by J.W. Emsley, Reidel, Dordrecht, Holland (1985), Ch. 3, and in *Recent Advances in Liquid Crystalline Polymers*, ed. by L.L. Chapoy, Elsevier, London and New York (1985), Ch. 7.
33. A.J. Leadbetter, R.M. Richardson and C.N. Colling, *J. Phys. (Paris)*, **36**, C1-37 (1975).
34. J.E. Lydon and C.J. Coakley, *J. Phys. (Paris)*, **36**, C1-45 (1975).
35. B.W. van der Meer, F. Postma, A.J. Dekker and W.H. de Jeu, *Mol. Phys.*, **45**, 1227 (1982).
36. M.J. Bradshaw and E.P. Raynes, *Mol. Cryst. Liq. Cryst.*, **91**, 145 (1983).
37. G.W. Gray, 'Molecular geometry and the properties of non-amphiphilic liquid crystals', *Adv. Liq. Cryst.*, **2**, 39 (1976).
38. M.A. Osman, *Z. Naturforsch.*, **38A**, 693 (1983).
39. M.A. Osman, *Z. Naturforsch.*, **38A**, 779 (1983).
40. M.A. Osman, *Mol. Cryst. Liq. Cryst.*, **128**, 45 (1985).
41. G.W. Gray, K.J. Harrison and J.A. Nash, *Electron. Lett.*, **9**, 130 (1973).
42. M. Schadt and W. Helfrich, *Appl. Phys. Lett.*, **18**, 127 (1971).
43. G.W. Gray, *J. Phys. (Paris)*, **36**, C1-337 (1975).
44. C. Hilsum, 'The anatomy of a discovery — Biphenyl liquid crystals', in *Technology of Chemicals and Materials for Electronics*, ed. by E.R. Howells, Ellis Horwood, Chichester, England (1984), p. 43.

45. G.W. Gray, 'Advances in liquid crystal materials for applications', BDH Special Publication, BDH Chemicals Ltd, Poole, England (1978) and BDH Chemicals Ltd Product Information on Liquid Crystal Materials.
46. M.A. Osman, *Z. Naturforsch.*, **34B**, 1092 (1979).
47. M.A. Osman and L.Revesz. *Mol. Cryst. Liq. Cryst.*, **56**, 133 (1980).
48. J. Ruoliene, P. Adomenas, A. Tubelyte, V. Seskauskas and V.A. Grozhik, *Mol. Cryst. Liq. Cryst.*, **78**, 211 (1981).
49. Yu. A. Fialkov, S.V. Shelyazhenko and L.M. Yagupol'skii, *Zh. Org. Khim.*, **19**, 1048 (1983); English translation p. 933.
50. W. Weissflog, *Z. Chem.*, **21**, 35 (1981).
51. H. Schubert, R. Dehne and V. Uhlig, *Z. Chem.*, **12**, 219 (1972).
52. H.-J. Deutscher, F. Kuschel, S. König, H. Kresse, D. Pfeiffer, A. Wiegeleben, J. Wulf and D. Demus, *Z. Chem.*, **17**, 64 (1977).
53. D. Demus in *Nonemissive Electrooptic Displays*, ed. by A.R. Kmetz and F.K. von Willisen, Plenum Press, New York (1976), p. 83; Brown, Boveri Symposium on Nonemissive Electrooptic Displays, Baden, Switzerland (1975).
54. H.-J. Deutscher, B. Laaser, W. Dölling and H. Schubert, *J. Prakt. Chem.*, **320**, 191 (1978).
55. M.J.S. Dewar and R.S. Goldberg, *J. Am. Chem. Soc.*, **92**, 1582 (1970).
56. M.J.S. Dewar and R.S. Goldberg, *J. Org. Chem.*, **35**, 2711 (1970).
57. M.J.S. Dewar and R.M. Riddle, *J. Am. Chem. Soc.*, **97**, 6658 (1975).
58. M.J.S. Dewar and A.C. Griffin, *J. Am. Chem. Soc.*, **97**, 6662 (1975).
59. R. Eidenschink, D. Erdmann, J. Krause and L. Pohl, *Angew. Chem., Int. Ed. Engl.*, **16**, 100 (1977).
60. R. Eidenschink, D. Erdmann, J. Krause and L. Pohl, *Angew. Chem., Int. Ed. Engl.*, **17**, 133 (1978).
61. A. Villiger, A. Boller and M. Schadt, *Z. Naturforsch.*, **34B**, 1535 (1979).
62. H. Minas, H.-R. Murawski, H. Stegemeyer and W. Sucrow, *J. Chem. Soc., Chem. Commun.*, **1982**, 308.
63. W. Sucrow, H. Minas, H. Stegemeyer, P. Geschwinder, H.-R. Murawski and C. Krüger, *Chem. Ber.*, **118**, 3332 (1985).
64. L. Pohl, R. Eidenschink, J. Krause and D. Erdmann, *Phys. Lett.*, **60A**, 421 (1977).
65. R. Eidenschink, J. Krause and L. Pohl, West Ger. Pat. 2,636,684 (1978); *Chem. Abs.*, **88**, 169762c (1978).
66. T. Szczucinski and R. Dabrowski, *Biul. Wojsk. Acad. Tech.*, **30**, 109 (1981).
67. L. Pohl, R. Eidenschink, J. Krause and G. Weber, *Phys. Lett.*, **65A**, 169 (1978).
68. M.J. Bradshaw and E.P. Raynes, *Mol. Cryst. Liq. Cryst.*, **72**, 35 (1981).
69. G.J. Brownsey and A.J. Leadbetter, *J. Phys. (Paris) Lett.*, **42**, L135 (1981).
70. D.A. Dunmur and A.E. Tomes, *Mol. Cryst. Liq. Cryst.*, **97**, 241 (1983).
71. G.W. Gray and S.M. Kelly, *J. Chem. Soc., Perkin Trans. 2*, **1981**, 26.
72. G.W. Gray, Abstract 12, Freiburger Arbeitstagung (1982).
73. N. Carr, G.W. Gray and D.G. McDonnell, *Mol. Cryst. Liq. Cryst.*, **97**, 13 (1983).
74. A. Boller, M. Cereghetti, M. Schadt and H. Scherrer, *Mol. Cryst. Liq. Cryst.*, **42**, 215 (1977).
75. D. Demus and H. Zaschke, *Mol. Cryst. Liq. Cryst.*, **63**, 129 (1981).
76. Y. Haramoto and H. Kamogawa, *Chem. Lett.*, **1985**, 79.
77. M. Schadt, M. Petrzilka, P.R. Gerber and A. Villiger, *Mol. Cryst. Liq. Cryst.*, **122**, 241 (1985).
78. M. Petrzilka, *Mol. Cryst. Liq. Cryst.*, **131**, 109 (1985).
79. M. Petrzilka and A. Germann, *Mol. Cryst. Liq. Cryst.*, **133**, 85 (1986).
80. L.K.M. Chan, P.A. Gemmell, G.W. Gray, D. Lacey and K.J. Toyne, to be published.
81. G.W. Gray and S.M. Kelly, *Angew, Chem., Int. Ed. Engl.*, **20**, 393 (1981).
82. N. Carr, G.W. Gray and S.M. Kelly, *Mol. Cryst. Liq. Cryst.*, **66**, 267 (1981).
83. G.W. Gray and D. Lacey, *Mol. Cryst. Liq. Cryst.*, **99**, 123 (1983).
84. U. Lauk, P. Skrabal and H. Zollinger, *Helv. Chim. Acta*, **64**, 1847 (1981).
85. U. Lauk, P. Skrabal and H. Zollinger, *Helv. Chim. Acta*, **66**, 1574 (1983).
86. M. Petrzilka and K. Schleich, *Helv. Chim. Acta*, **65**, 1242 (1982).
87. M. Cereghetti, R. Marbet and K. Schleich, *Helv. Chim. Acta*, **65**, 1318 (1982).
88. H. Sato, H. Takatsu, K. Takeuchi, Y. Fujita, M. Tazume and H. Ohnishi, Jpn. Pat. 118,526 (1982); *Chem. Abs.*, **98**, 34305s (1983); see reference [38].
89. K. Praefcke, D. Schmidt and G. Heppke, *Chem.-Ztg.*, **104**, 269 (1980).

90. V. Reiffenrath and F. Schneider, *Z. Naturforsch.*, **36A**, 1006 (1981).
91. H.M. Abdullah, MSc thesis, University of Hull (1983).
92. H.M. Abdullah, G.W. Gray and K.J. Toyne, *Mol. Cryst. Liq. Cryst.*, **124**, 105 (1985).
93. H. Schubert, H.-J. Lorenz, R. Hoffmann and F. Franke, *Z. Chem.*, **6**, 337 (1966).
94. D. Demus, H. Demus and H. Zaschke, *Flüssige Kristalle in Tabellen*, VEB Deutscher Verlag für Grundstoffindustrie, Leipzig, German Democratic Republic (1974), p. 230.
95. M.A. Osman, and T. Huynh-Ba, *Mol. Cryst. Liq. Cryst.*, **116**, 141 (1984).
96. H. Schubert and V. Heise, unpublished result (1978); see reference [3], p. 270.
97. Jpn. Pat. application, 125,518 (1983) and Jpn. Pat. 179,293 (1982); thanks are expressed to Dr Eidenschink for providing these references.
98. G.W. Gray, *Mol. Cryst. Liq. Cryst.*, **63**, 3 (1981).
99. R. Eidenschink, *Kontakte (Merck-Darmstadt)*, **3/80**, 12 (1980).
100. G.W. Gray, K.J. Harrison and J.A. Nash, *J. Chem. Soc., Chem. Commun.*, **1974**, 431.
101. R. Eidenschink, J. Krause and L. Pohl, Ger. Pat. 2,701,591 (1978); *Chem. Abs.*, **89**, 215085e (1978).
102. D. Coates and G.W. Gray, *Mol. Cryst. Liq. Cryst.*, **37**, 249 (1976).
103. A. Boller and H. Scherrer, Ger. Pat. 2,415,929 (1974); *Chem. Abs.*, **82**, 24776b (1975).
104. G.W. Gray and D.G. McDonnell, *Mol. Cryst. Liq. Cryst.*, **53**, 147 (1979).
105. H.J. Deutscher, C. Seidel, H. Zaschke, D. Pfeiffer, F. Kuschel and D. Demus, East Ger. Pats. 137, 597 and 137,928 (1978); *Chem. Abs.*, **92**, 155917c (1980) and **92**, 189258d (1980) respectively.
106. G.W. Gray and S.M. Kelly, *J. Chem. Soc., Chem. Commun.*, **1979**, 974.
107. M.A. Osman and T. Huynh-Ba, *Helv. Chim. Acta*, **67**, 959 (1984).
108. J.A. Hirsch in *Topics in Stereochemistry*, ed. by N.L. Allinger and E.L. Eliel, Vol. 1, Interscience, New York (1967), p. 199.
109. G.W. Gray, N.A. Langley and K.J. Toyne, *Mol. Cryst. Liq. Cryst.*, **64**, 239 (1981).
110. G.W. Gray, N.A. Langley and K.J. Toyne, *Mol. Cryst. Liq. Cryst.*, **98**, 425 (1983).
111. D.J. Byron, D.A. Keating, M.T. O'Neill, R.C. Wilson, J.W. Goodby and G.W. Gray, *Mol. Cryst. Liq. Cryst.*, **58**, 179 (1980).
112. P.J. Bullock, D.J. Byron, D.J. Harwood, R.C. Wilson and A.M. Woodward, *J. Chem. Soc., Perkin Trans. 2*, **1984**, 2121.
113. D.J. Byron, R.C. Wilson, P.A. Baker, I. Danilewicz, G.S. Hayer, D.A. Hewison, J.M. Taylor and J.M. Wyer, *J. Chem. Soc., Perkin Trans. 2*, **1985**, 297.
114. N.A. Vaz, S.L. Arora, J.W. Doane and A. de Vries, *Mol. Cryst. Liq. Cryst.*, **128**, 23 (1985).
115. G.W. Gray and A. Mosley, *Mol. Cryst. Liq. Cryst.*, **37**, 213 (1976).
116. R.J. Cox and N.J. Clecak, *Mol. Cryst. Liq. Cryst.*, **37**, 241 (1976).
117. J. Malthete, J. Canceill, J. Gabard and J. Jacques, *Tetrahedron*, **37**, 2815 (1981).
118. R.J. Cox and N.J. Clecak, *Mol. Cryst. Liq. Cryst.*, **37**, 263 (1976).
119. J.C. Dubois, A. Zann and N.H. Tinh, *C.R. Hebd. Seances Acad. Sci., Ser. C*, **284**, 137 (1977).
120. R. Steinsträsser, *Z. Naturforsch.*, **27B**, 774 (1972).
121. D. Coates and G.W. Gray, *Mol. Cryst. Liq. Cryst.*, **41**, 197 (1978).
122. S. Richter and H.-J. Deutscher. *Z. Chem.*, **23**, 21 (1983).
123. M.A. Osman and L. Revesz, *Mol. Cryst. Liq. Cryst.*, **56**, 105 (1979) and **56**, 157 (1980); these two references are the same paper.
124. G.W. Gray and S.M. Kelly, *J. Chem. Soc., Chem. Commun.*, **1980**, 465.
125. G.W. Gray and S.M. Kelly, *Mol. Cryst. Liq. Cryst.*, **75**, 95 (1981).
126. G.W. Gray and S.M. Kelly, *Mol. Cryst. Liq. Cryst.*, **95**, 101 (1983).
127. D. Coates and G.W. Gray, *Mol. Cryst. Liq. Cryst.*, **41**, 119 (1978).
128. H.-J. Deutscher, C. Seidel, M. Körber and H. Schubert, *J. Prakt. Chem.*, **321**, 47 (1979).
129. M. Petrzilka and M. Schadt, Ger. Pat. 3,237,367 (1981); *Chem. Abs.*, **99**, 222510z (1983).
130. B.M. Andrews, G.W. Gray and M.J. Bradshaw, *Mol. Cryst. Liq. Cryst.*, **123**, 257 (1985).
131. S.M. Kelly and Hp. Schad, *Helv. Chim. Acta*, **68**, 1444 (1985).
132. N. Carr and G.W. Gray, *Mol. Cryst. Liq. Cryst.*, **124**, 27 (1985).
133. J.E. Fearon, G.W. Gray, A.D. Ifill and K.J. Toyne, *Mol. Cryst. Liq. Cryst.*, **124**, 89 (1985).
134. R. Eidenschink, *Mol. Cryst. Liq. Cryst.*, **123**, 57 (1985).
135. R. Eidenschink, M. Römer, G. Weber, G.W. Gray and K.J. Toyne, Ger. Pat. 3,321,373 A1 (1983); *Chem. Abs.*, **102**, 166362u (1985).
136. M.A. Osman, *Mol. Cryst. Liq. Cryst.*, **72**, 291 (1982).

137. W. Sucrow and H. Wolter, *Chem. Ber.*, **118**, 3350 (1985).
138. W. Weissflog and D. Demus, *Cryst. Res. Technol.*, **18**, K21 (1983) and **19**, 55 (1984).
139. W. Weissflog and D. Demus, *Mol. Cryst. Liq. Cryst.*, **129**, 235 (1985).
140. A. Hauser, R. Rettig, Ch. Selbmann, W. Weissflog, J. Wulf and D. Demus, *Cryst. Res. Technol.*, **19**, 261 (1984).
141. D. Demus, A. Hauser, Ch. Selbmann and W. Weissflog, *Cryst. Res. Technol.*, **19**, 271 (1984).
142. H.-J. Deutscher, M. Körber, H. Altmann and H. Schubert, *J. Prakt. Chem.*, **321**, 969 (1979).
143. D.J. Byron, G.W. Gray, A. Ibbotson and B.M. Worrall, *J. Chem. Soc.*, **1963**, 2246.
144. P. Balkwill, D. Bishop, A. Pearson and I. Sage, *Mol. Cryst. Liq. Cryst.*, **123**, 1 (1985).
145. M.R.C. Gerstenberger and A. Haas, *Angew. Chem., Int. Ed. Engl.*, **20**, 647 (1981).
146. L.K.M. Chan, G.W. Gray and D. Lacey, *Mol. Cryst. Liq. Cryst.*, **123**, 185 (1985).
147. L.D. Field and S. Sternhell, *J. Am. Chem. Soc.*, **103**, 738 (1981).
148. S.J. Branch, D.J. Byron, G.W. Gray, A. Ibbotson and B.M. Worrall, *J. Chem. Soc.*, **1964**, 3279.
149. D.J. Byron, G.W. Gray and B.M. Worrall, *J. Chem. Soc.*, **1965**, 3706.
150. G.W. Gray, J.B. Hartley, A. Ibbotson and B. Jones, *J. Chem. Soc.*, **1955**, 4359.
151. G.W. Gray and B.M. Worrall, *J. Chem. Soc.*, **1959**, 1545.
152. G.W. Gray and S.M. Kelly, *Mol. Cryst. Liq. Cryst.*, **75**, 109 (1981).
153. S.M. Kelly, *J. Chem. Soc., Chem. Commun.*, **1983**, 366.
154. S.M. Kelly and Hp. Schad, *Helv. Chim. Acta*, **67**, 1580 (1984).
155. G.W. Gray, C. Hogg and D. Lacey, *Mol. Cryst. Liq. Cryst.*, **67**, 1 (1981).
156. M.E. Neubert, L.T. Carlino, D.L. Fishel and R.M. D'Sidocky, *Mol. Cryst. Liq. Cryst.*, **59**, 253 (1980).
157. G.W. Gray and B. Jones, *J. Chem. Soc.*, **1955**, 236.
158. Hp. Schad and M.A. Osman, *J. Chem. Phys.*, **75**, 880 (1981).
159. M.A. Osman and T. Huynh-Ba, *Mol. Cryst. Liq. Cryst.*, **82**, 331 (1983).
160. M.A. Osman and T. Huynh-Ba, *Z. Naturforsch.*, **38B**, 1221 (1983).
161. Hp. Schad and M.A. Osman, *J. Chem. Phys.*, **79**, 5710 (1983).
162. Hp. Schad and S.M. Kelly, *J. Chem. Phys.*, **81**, 1514 (1984).
163. G. Heppke and E.-J. Richter, *Z. Naturforsch.*, **33A**, 185 (1978).
164. B. Engelen and F. Schneider, *Z. Naturforsch.*, **33A**, 1077 (1978).
165. B. Engelen, G. Heppke, R. Hopf and F. Schneider, *Ann. Phys.*, **3**, 403 (1978).
166. S.M. Kelly, *Helv. Chim. Acta*, **67**, 1572 (1984).
167. M. Sasaki, K. Takeuchi, H. Sato and H. Takatsu, *Mol. Cryst. Liq. Cryst.*, **109**, 169 (1984).
168. K. Toriyama and D.A. Dunmur, *Mol. Phys.*, **56**, 479 (1985).
169. D.G. McDonnell, E.P. Raynes and R.A. Smith, *Mol. Cryst. Liq. Cryst.*, **123**, 169 (1985).
170. R. Eidenschink, G. Haas, M. Römer and B.S. Scheuble, *Angew. Chem., Int. Ed. Engl.*, **23**, 147 (1984).
171. H. Schubert and W. Schulze, unpublished results (1968); see reference [2], p. 34.
172. R. Steinsträsser and F. del Pino, Ger. Pat. 2,450,088 (1974); *Chem. Abs.*, **85**, 46219m (1976).
173. X.J. Hong, M.J. Ge, X.M. Zhao, Z.R. Fen and Z.J. Liu, *Mol. Cryst. Liq. Cryst.*, **99**, 81 (1983).
174. S.M. Kelly and T. Huynh-Ba, *Helv. Chim. Acta*, **66**, 1850 (1983).
175. M.A. Osman and T. Huynh-Ba, *Mol. Cryst. Liq. Cryst.*, **92**, 57 (1983).
176. M. Schadt, K. Schleich, G. Trickes and M. Petrzilka, Br. Pat. 2,102,803A (1982); West Ger. Pat. 3,228,350 (1981); *Chem. Abs.*, **99**, 30802z (1983).
177. M. Schadt, M. Petrzilka, P.R. Gerber, A. Villiger and G. Trickes, *Mol. Cryst. Liq. Cryst.*, **94**, 139 (1983).
178. M.P. Burrow, G.W. Gray, D. Lacey and K.J. Toyne, *Z. Chem.*, **26**, 21 (1986).
179. R. Eidenschink, *Mol. Cryst. Liq. Cryst.*, **94**, 119 (1983).
180. G.W. Gray, D. Lacey, J.E. Stanton and K.J. Toyne, *Liq. Cryst.*, **1**, 407 (1986).
181. G.W. Gray and S.M. Kelly, *Mol. Cryst. Liq. Cryst.*, **104**, 335 (1984).
182. G.W. Gray and K.J. Harrison, *Mol. Cryst. Liq. Cryst.*, **13**, 37 (1971).
183. G.W. Gray and D.G. McDonnell, *Mol. Cryst. Liq. Cryst.*, **34**, 211 (1977).
184. W. Weissflog, A. Wiegeleben and D. Demus, *Mater. Chem. Phys.*, **12**, 461 (1985).
185. G. Solladié and R.G. Zimmermann, *Angew. Chem., Int. Ed. Engl.*, **24**, 64 (1985).
186. K. Seto, S. Takahashi and T. Tahara, *J. Chem. Soc., Chem. Commun.*, **1985**, 122.

187. Y. Haramoto, A. Nobe and H. Kamogawa, *Bull. Chem. Soc. Jpn.*, **57**, 1966 (1984).
188. Y. Haramoto, K. Akazawa and H. Kamogawa, *Bull. Chem. Soc. Jpn.*, **57**, 3173 (1984).
189. D.J. Byron, D. Lacey and R.C. Wilson in *Liquid Crystals and Ordered Fluids*, Vol. 4, ed. by J.F. Johnson and A.C. Griffin, Plenum Press, New York (1984), p. 745.

3 Materials requirements for nematic and chiral nematic electrooptical displays

I. Sage
BDH Limited

3.1 Introduction

The requirement of the electrooptic display industry for improved liquid crystal (LC) materials for device development and manufacture has provided much of the impetus for LC research over the last two decades. Conversely, it is the success of chemists in synthesizing novel materials which approach the target specifications set by device engineers that has made possible the modern passive electrooptic display industry. Historically, the discovery of optical effects in liquid crystals[1–3] predated the availability of stable room temperature mesophases to exploit them, and this problem[4–6] was only fully resolved by the synthesis of cyanobiphenyls[7–9] in the early 1970s. Once these materials became available, however, the aspirations of device producers progressively expanded to encompass devices displaying more information at higher density and over wider ranges of environmental conditions, which place increasingly stringent demands on the physical characteristics of the LC fluid used. Today the development of liquid crystals for displays is a mature science and the ideal material for a given application might be characterized by target values for a dozen physical properties.

It remains true that the targets set for LC materials by users have always outstripped the ability of chemists to supply them, and probably exceed in several respects what can be achieved in real molecules. Nevertheless, sufficient progress has been made towards such targets to extend the range of applicability of LC displays far beyond their original uses in watches and calculators. Incremental improvements in materials are likely to continue, particularly in fluids for the established twisted nematic display mode, until a considerably superior device technology is discovered.

The problem of developing improved LC materials for devices may be considered in two parts. Firstly, an understanding of the operation of the device is necessary to deduce which physical properties of the fluid will affect its performance. Such an understanding must include the effect of an applied electric field or current on the alignment of the LC, the change in optical properties of the LC cell resulting from this realignment and possibly the psychophysical aspects which relate these optical phenomena to a perceived change of contrast in a display. These questions are largely the domain of the device physicist and in an ideal situation lead to the target values for LC physical properties referred to above. Secondly, it is necessary to develop an understanding of the structure-property relations which exist in LC single materials and the composition-property relations pertaining in mixtures in order to design fluids that approach these targets. It must be admitted from the outset, however, that although sufficient progress has been made regarding the first of these problems to provide useful guidelines to materials scientists, our understanding of how given physical properties may be obtained in useful combinations in practical liquid crystals is far from perfect. Much of the groundwork in the field was carried out even before device modelling was well established, and most progress in LC formulation has probably been made by detailed empirical experimentation, with the results being rationalized and refined subsequently following the development of the necessary theory.

All LC display devices produced commercially contain fluids which are mixtures of several or many LC components. Overall, however, the number of classes of compounds which have found commercial usefulness is quite small, and the number of structural subunits which are employed is even more limited (see Table 3.1). This fact arises naturally from the constraints imposed by practical applications: the fluids employed in devices must be stable, colourless, mesomorphic over a wide range in the room temperature region and operate at low voltage and power levels. The first two of these criteria alone eliminate a vast

Table 3.1 Some technologically important nematic liquid crystals

	N–I	Δn	$\Delta\varepsilon$	k_{33}/k_{11}
R–(phenyl)–(phenyl)–CN	35	0.18	11.5	1.3
R–(cyclohexyl)–(phenyl)–CN	50	0.1	9.7	1.6
R–(pyrimidinyl)–(phenyl)–CN	50	0.18	19.7	1.2
R–(phenyl)–CO_2–(phenyl)–CN	55	0.15	19.7	1.7
R–(phenyl)–CO_2–(phenyl)–R′	20	0.13	0.5	1.2
R–(cyclohexyl)–CO_2–(phenyl)–OR′	75	0.07	−1.0	1.3
R–(cyclohexyl)–(cyclohexyl)–CN	85	0.06	4.4	1.5
R–(phenyl)–CO_2–(fluorophenyl, F)–CN	30	0.14	48.9	1.7
R–(1,3-dioxanyl)–(phenyl)–CN	50	0.09	13.3	1.4

number of LC compounds, and it is a measure of progress in the subject that no further consideration will be given to them.

The plan of this chapter is to consider first the physical properties of liquid crystals which are most important for display operation, with some mention of their relation to molecular structure and behaviour in mixtures. The more important display effects are then considered in turn, and any steps necessary to optimize LC mixtures for each are touched upon.

3.2 Nematic LC properties

3.2.1 Nematic phase range and order parameter

The nematic liquid crystal (NLC) phase is commonly bounded by a smectic (S) or crystalline phase at low temperature and an isotropic liquid at high temperature; the correlation between molecular structure, phase sequence and transition temperatures is considered in Chapter 2. For the purposes of device applications, an N phase range between −10 and +60 °C represents the *most basic* requirement, and this in itself dictates the necessity for mixtures to be used. The resolution of the problem arises from the different behaviour of clearing (T_c) and melting (T_m) temperatures in mixtures as a function of composition:

$$T_c = \sum_i a_i T_{ci} \tag{3.1}$$

where a_i is the percentage of component i in the mixture and T_{ci} its clearing point. The solubility of components in a mixture which behaves as a thermodynamically ideal system follows the van't Hoff relation[10–12], which is conveniently recast into the Schroeder–Van Laar equation:

$$\ln[C]_i = \frac{\Delta H_{fi}}{R}\left[\frac{1}{T_{mi}} - \frac{1}{T}\right] \tag{3.2}$$

where $[C]_i$ is the solubility of the component at temperature T, ΔH_{fi} its latent heat of fusion or solution and T_{mi} its melting point. For a given set of components there is a unique eutectic temperature T_e at which

$$\sum_i [C]_i = 1 \tag{3.3}$$

and the corresponding composition of mixture melts sharply at T_e, which is a lower temperature than the melting point of any of the components (see Figure 3.1). In principle, low melting LC mixtures can be obtained from a knowledge of the melting temperatures and enthalpies of the components by solution of equations 3.2 and 3.3, but, in practice, deviations from ideal behaviour are sufficiently common and large to make this approach unreliable[13]. Alternative procedures have been suggested[14] for the experimental determination of eutectic points.

The NLC phase is distinguished from an isotropic liquid by the long-range ordering of rod-like molecules; this is never very perfect, but it is the basis of the observed anisotropy[15, 16] of the various physical properties of N phases. It is convenient to define an order parameter which on the one hand reflects the ordering and symmetry of the phase on a molecular level, while on the other hand providing a scaling factor for the observed physical properties of the bulk liquid. It is the latter role which is paramount for the discussion of display-related materials, for it is the anisotropy, particularly of optical and dielectric properties, that makes LCs into useful display components.

The director of an NLC is defined as a unit vector pointing along the preferred local alignment direction of the molecular long axes. N liquids behave as uniaxial materials in which many physical properties display two principal values, measured respectively parallel and perpendicular to the director, the anisotropy of the property also being defined as the difference between these principal values. All observed properties of NLC are invariant under reversal of the director, i.e. N molecules with different end groups have an equal probability of pointing either one in a given direction. Now we require a scalar quantity that describes the average alignment of molecules with the director, incorporates this

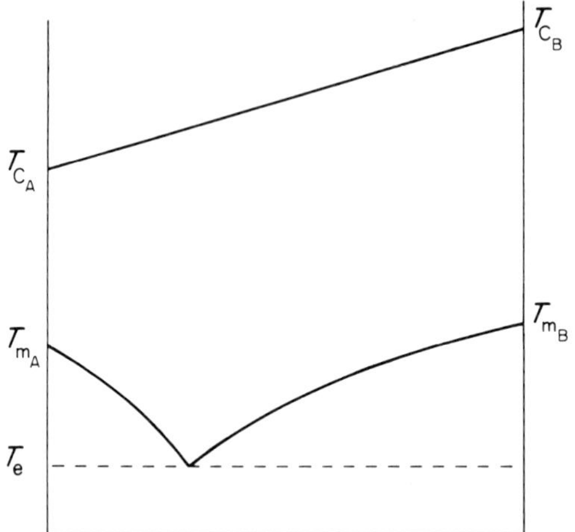

Figure 3.1 Expansion of nematic phase range in LC eutectic mixtures.

symmetry, is zero valued for an isotropic liquid and has the value unity for a (hypothetical) perfectly aligned NLC.

The simplest such order parameter is

$$S = \tfrac{1}{2}\langle 3\cos^2\theta - 1\rangle \tag{3.4}$$

where θ is the angle between the director and the long axis of each molecule, the molecules being regarded as cylindrically symmetric. The brackets denote an average over the ensemble of molecules.

The theoretical treatment of N ordering has been a subject of intense research and has achieved some notable success, but in common with the theory of almost all LC properties it has not reached the point where quantitative predictions of order parameter or qualitative trends in structure-property relations can be derived. The relatively simple Maier–Saupe theory[17, 18] gives good semiquantitative agreement with experiment by regarding NLC molecules as cylinders which interact only by long-range[19] dispersion forces. It predicts a universal behaviour of S with temperature which is, unfortunately, not available as an analytic expression, but only as a self-consistency relation. The Maier–Saupe order parameter can be expressed approximately[16] by the equation:

$$S \simeq \left(1 - \frac{yT}{T_{N-I}}\right)^z \tag{3.5}$$

where $y \simeq 0.98$ and $z \simeq 0.22$, while for a series of experimental data[20], most materials gave best-fit parameters of $y = 0.98$–0.99 and $z = 0.14$–0.18. Many attempts have been made to extend the Maier–Saupe approach or to begin from alternative premises; there is a considerable cost in increased complexity before any improvement in experimental concordance[21] is obtained.

The experimental determination of order parameter is normally carried out by measuring an appropriate physical quantity proportional to S. In order to obtain absolute values, the property must be known in the perfectly aligned state and this is achieved[22] by fitting data to a function such as equation 3.5 above and extrapolating to 0 K. The form of the fitting equation is important and alternatives[23] have been suggested. In a few cases, measurements made on the crystalline solid may be used as an alternative.

The most usual quantities chosen are

$$S \propto \frac{n_e^2 - n_o^2}{\bar{n}^2 - 1} \tag{3.6}$$

or

$$S \propto \chi_{\parallel} - \chi_{\perp} \tag{3.7}$$

The use of magnetic anisotropy in equation 3.7 leads to a high degree of confidence in the theoretical treatment of data, but it is difficult to make sufficiently accurate experimental measurements. Refractive indices may be measured accurately for use in equation 3.6 but their interpretation[22] is complicated by local field effects.

Few sets of data measured and analysed under comparable conditions are available[20, 22, 23] to establish structure-property relations in the behaviour of order parameters, and sets of data which do exist may contradict one another. The Maier–Saupe theory predicts that an increase in clearing point should increase S at fixed temperature, and this is borne out; such consistent experimental data as do exist indicate that systems based on cyclohexane and bicyclo[2,2,2]octane rings show higher ordering than aromatic systems, and cyano-terminated molecules give higher S values than those carrying non-polar terminal groups. The use of more advanced techniques such as NMR[15, 24] allows the measurement of S for individual components of a mixture, or even for different structural groups within a flexible molecule.

For displays based on LCs containing dichroic dyes, the order parameter derived from the anisotropy of the extinction coefficient is of particular importance. A rod-like dye molecule tends to adopt the ordering of a nematic host in which it is dissolved, and many such dyes have their transition moment approximately parallel to the long axis. Then the order parameter of such a dye is defined by

$$S = \frac{A_{\parallel} - A_{\perp}}{A_{\parallel} + 2A_{\perp}} \tag{3.8}$$

and this definition is also used in cases where the transition moment does not coincide with the long axis or where the very definition of a molecular long axis becomes problematic. Defined in this way, the order parameter is a function[25–27] both of the dyestuff and the host solvent. The order parameter may be higher for the dye than for the host and may be correlated with the molecular structure of the dye in similar terms to the correlation of clearing point for a nematogen. It is notable, however, that dyes based[28] on anthraquinone which do not conspicuously resemble long, thin rods may give high order parameters. Dye order parameters have been widely quoted in literature sources[26–33].

3.2.2 Dielectric constants

The electrical capacitance of two parallel conducting plates is a function of the medium between them, and this fact is used to define the dielectric constants of an LC:

$$\varepsilon_{\|,\perp} = \frac{C_{\|,\perp}}{C_c} \tag{3.9}$$

A cell is first measured empty to give C_c; it is then filled with LC and measured with a suitable field applied to align the director respectively parallel or perpendicular to the sensing electric field to provide values of $C_{\|}$ and $C_{\perp}$. It is usual to discuss the dielectric anisotropy of an LC material as defined by

$$\Delta\varepsilon = \varepsilon_{\|} - \varepsilon_{\perp} \tag{3.10}$$

For electrooptic display devices, $\Delta\varepsilon$ is a quantity of paramount importance, as its magnitude directly determines the strength of the interaction between an LC and an applied electric field. Its sign also determines the geometry of display devices; under a large applied field the LC director aligns so that the larger dielectric constant lies parallel to the field.

A predictively useful insight into the relation between dielectric constants and molecular structure is provided by the theory of Maier and Meier[34]. Considering mesogenic molecules as centres of anisotropic polarizability with point dipoles, equations are derived for dielectric constants in terms of molecular properties and order parameter, yielding

$$\Delta\varepsilon \propto S[\Delta\alpha - C\mu^2(1 - 3\cos^2\beta)] \tag{3.11}$$

where $\Delta\alpha$ is the polarizability anisotropy, μ the dipole moment and β the angle between the molecular long axis and the dipole moment. If $\beta \simeq 54.7°$, the second term in brackets becomes zero and a small positive $\Delta\varepsilon$ is expected due to the positive $\Delta\alpha$ of the molecules. For other values of β, the dipole moment makes a contribution and the constant C is such that it often dominates the $\Delta\alpha$ term. Design of molecules with a given $\Delta\varepsilon$ would thus appear to be dependent on the introduction of a suitable dipole moment parallel or perpendicular to the molecular long axis. Although this is largely true, there are factors which complicate the true situation.

There is now very good evidence[35] that mesogenic molecules bearing strongly dipolar terminal groups form associated pairs on a transitory basis. Both head-to-head and head-to-tail pairing occurs[36, 112], but antiparallel association predominates and reduces the effective molecular dipole moment. The degree of association is a function of molecular structure, but is not readily predicted. The problems encountered in obtaining negative $\Delta\varepsilon$ materials are more prosaic. It will be noted from the term in β in equation 3.11 that a lateral dipole moment is only one half as effective in inducing negative anisotropy as a terminal one is in obtaining positive behaviour. This is because there are two axes perpendicular to the LC director along which the dipole may lie, only one of which may be parallel to the electric field. Furthermore in an LC system based broadly on hexagonal ring systems, it is difficult to obtain a β close to 90°. Finally, a strong dipole moment is usually resident in a substituent group[37–39], the lateral disposition of which may adversely affect LC phase stability or drastically raise the viscosity of the system. Important advances have been made in this area using heterocyclic systems[40] and axial-substituted cyclohexane derivatives[41], but it remains the case that strongly positive materials are more readily obtained than strongly negative ones.

In mixtures, a nearly linear variation of dielectric constants with composition is expected, except in mixtures of terminal polar/non-polar mesogens[42, 43], when disruption of the pairing referred to above causes large deviations from ideal behaviour.

3.2.3 Elastic constants

In LC electrooptic devices, it is the operation of an electric field on the dielectric anisotropy of the fluid that tends to turn on the cell, while the turn-off is usually driven by elastic forces. The elastic behaviour of LCs is very weak when compared to that of solids and is difficult to detect by mechanical means[44], but is sufficient to drive the realignment of the LC director in a time scale of the order of milliseconds. Elasticity in LCs is not dimensionally equivalent to that of solids: all deformations in LCs are curvatures and can be expressed as a combination of the basic operations of splay, twist and bend of the director, mediated by elastic constants k_{11}, k_{22} and k_{33} respectively, whose units are those of force (Figure 3.2). Although the absolute values of elastic constants help to determine the relaxation speed of display devices, it is ratios of the elastic constants, and particularly the ratio k_{33}/k_{11}, that have received most attention because of their effect on the shape of the electrooptic switching curve in displays. This point will be returned to in Section 3.3.2 on the twisted nematic cell.

No useful theory exists for the prediction of elastic constants or their ratios. Although analyses have been performed, the results are at variance with experiment[45, 46], perhaps because only the grossest aspects of molecular structure are considered and it is necessary to rely on empirical rules based on experimental measurements. In general, a low value of k_{33}/k_{11}, which is desirable in many display modes, can best be obtained by using aromatic and heterocyclic ring systems, high members of homologous series, and by excluding lateral substituents (see Table 3.2). Absolute values of elastic constants appear

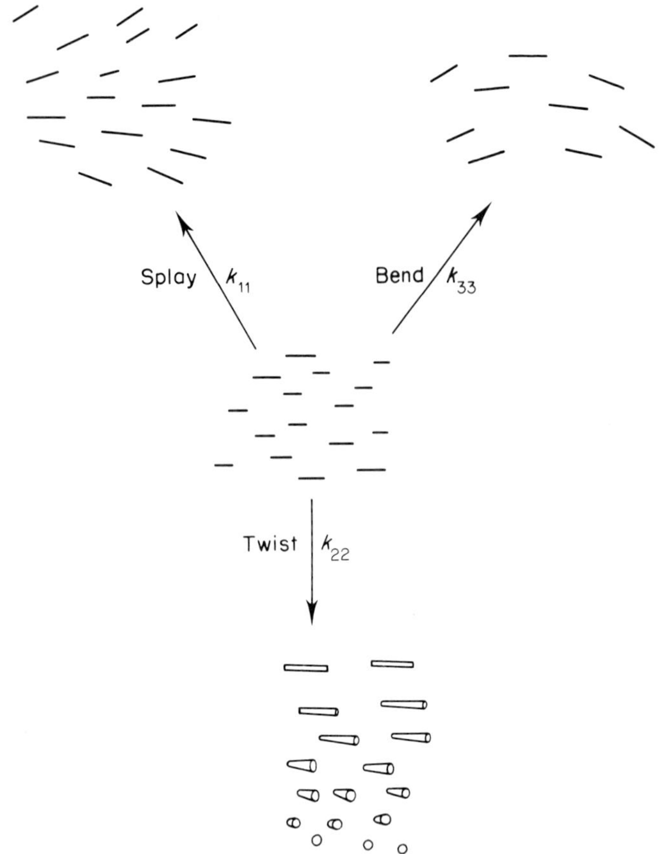

Figure 3.2 Three fundamental elastic distortions of nematic fluids.

to correlate approximately with clearing points, and only approximately follow the theoretical prediction that

$$k_{ii} \propto S^2 \tag{3.12}$$

The elastic constants k_{22} and k_{33} commonly diverge to large values on approaching a smectic phase.

In mixtures of components of broadly similar structure (particularly in regard to the presence of dipolar terminal groups), a linear variation of elastic constants with composition[47, 48] is observed. An almost linear variation of k_{11} and k_{22} has also been reported in mixtures of dissimilar components, but k_{33} and k_{33}/k_{11} deviate[45, 48] markedly from a linear law to give lower values than would be expected.

3.2.4 *Refractive indices*

The refractive indices of an LC are determined by the polarizability of the

Table 3.2 Correlation between k_{33}/k_{11} ratio and structure in LC subunits

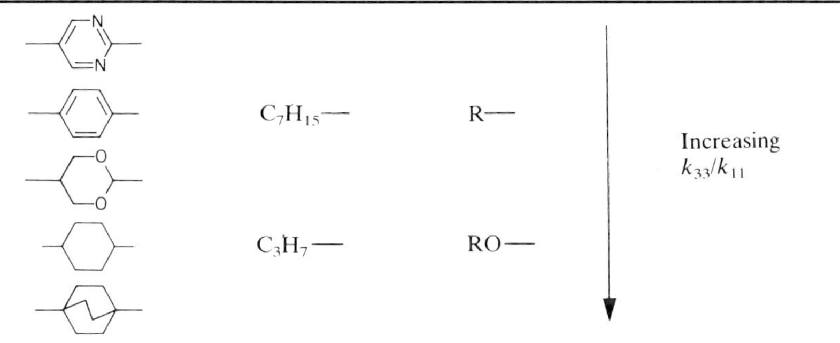

molecules at optical frequencies, and therefore depend largely on the extent of conjugated π bonding throughout the molecule. NLCs behave as optically positive uniaxial materials[49] and it is usual to refer to $n_{\|}$ and $n_{\perp}$ as n_e and n_o respectively and to define the birefringence as

$$\Delta n = n_e - n_o \tag{3.13}$$

Although the dependence of refractive indices on order parameter is most accurately represented by equation 3.6, for practical values and most purposes it may be taken that

$$\Delta n \propto S \tag{3.14}$$

Although the refractive indices of LCs can be treated fairly successfully by theory, it is relatively easy to estimate values for new materials by analogy with numerous published data for a wide range of LC systems. As indicated above, birefringence is largely determined by the presence of aromatic rings and π-bonded terminal or linking groups in the LC molecule. In mixtures, the assumption of a linear variation of Δn with composition is a useful approximation, although marked deviations can be observed in some systems.

3.2.5 Viscosity

The dynamic response of LC displays is limited[50] by the viscosity of the fluid. In an NLC, the viscosity displays anisotropy like other physical parameters, but both the theory and phenomenology of LC viscosities is rather complex. Five independent viscosities are required to characterize a nematic liquid: three of these (the Miesowicz viscosities) represent conventional shear flow with different dispositions of the director to the shear direction, while the other two include director rotation or coupling between the director and the flow pattern. Experimental determinations of these quantities are hindered by the tendency for shear to rotate the director, so very few complete sets of viscosity data exist for LCs. It is normal practice in the display materials industry[51, 52] to determine a capillary flow viscosity or other bulk viscosity, the significance of which is poorly characterized. Such figures correlate well with the response times of display devices within the variation to be expected from concomitant changes in elastic properties.

The bulk viscosity is not principally dependent on order parameter, although a discontinuity in the plot of η against temperature at the clearing point reveals that the effect of long-range order is significant. The dominant term in the temperature dependence[51] is a conventional Arrhenius-type behaviour:

(3.15)

As a general guide, η changes by a factor of 3–5 over 20 °C. Viscosity is minimized in molecular structures of minimal polarity and polarizability, with short terminal groups and lacking lateral substituents.

The viscosity of an N mixture can be estimated[53] by summing the logarithms of the viscosities of its components:

(3.16)

3.3 Display construction

In order to utilize the anisotropic properties of LCs in an electrooptic display, it is necessary to provide means whereby different regions of an LC layer will show different optical properties. The principles of cell construction[54] described in this section are broadly applicable to LC displays; the special requirements of different display modes for specific alignment geometry, polarization or LC properties will be mentioned later.

All of the common LC display modes employ the liquid crystal as a thin layer, typically 5–20 μm thick, supported between parallel glass plates (Figure 3.3). Each sheet of glass carries on its surface an aligning layer whose function is to hold the LC director in a defined configuration. In the absence of any voltage applied to the cell, the LC director throughout the display adopts a configuration dependent upon the alignment geometry, the elastic constant ratios of the fluid and the LC helical pitch, if applicable. Electric voltage or current is applied to different regions of the LC layer by means of transparent electrically conductive tin oxide or indium–tin oxide deposited on the glass, then etched into patterns which define the active areas of the display. The LC overlying these electrode areas behaves as a light shutter when the cell is actuated. Thin conductive tracks brought out to the edge of the display enable independent connections to be made to each segment of simple patterns such as the numeric seven-segment character (Figure 3.4), and activation of the segments in different combinations allows any digit to be represented. More complex patterns such as the 5×7 alphanumeric matrix must be driven by a multiplex scheme (see Section 3.3.2(a)), as it is not possible to make a separate connection to each pixel in the display. The edges of the display are sealed using either an epoxy-based adhesive or a low-melting glass frit. In large area displays, it is common practice to regulate the spacing of the

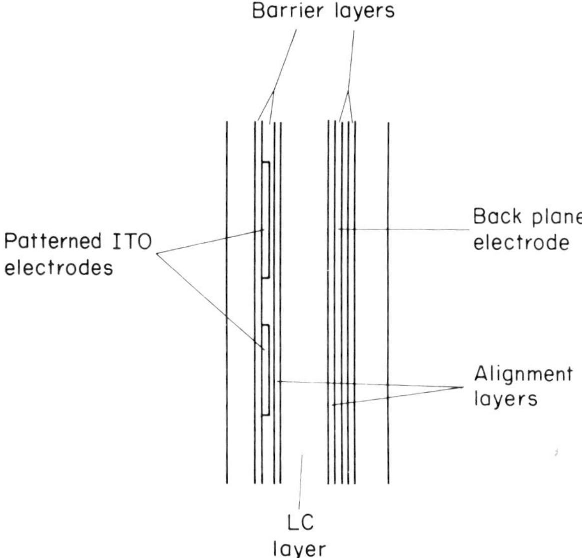

Figure 3.3 Liquid crystal display cell construction.

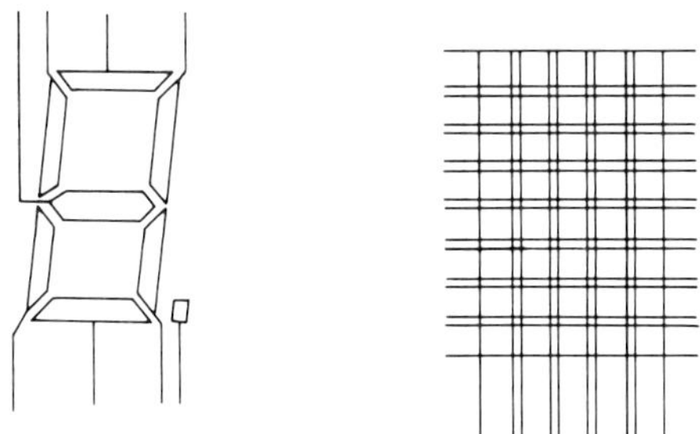

Figure 3.4 Electrode patterns for a directly driven seven-segment display and a 5 × 8 dot matrix character.

glass plates by incorporating a low density of fibres or spheres of closely defined diameter into the display area, but this is not usually necessary in common watch or calculator cells. Finally, one or more barrier layers may be present on the glass plates. These are intended to prevent the LC from experiencing any d.c. bias voltage by providing electrical insulation and to avoid contamination of the LC by ionic impurities in the glass.

3.3.1 Alignment methods

The ability of solid surfaces to orientate the director of an LC has long been recognized[55], and its exploitation and control are necessary to almost all LC optical devices, although the geometry required varies between display modes. The several alignment methods which are or have been important in display cells differ in their mode of application, but will here be categorized according to the angle formed between the director and the plane of the substrate.

(a) Planar alignment. Evaporation of a range of inorganic materials at angles of *ca.* 30° from the plane of the substrate[56] results in a surface coating that aligns the LC director in the plane of the substrate and perpendicular to the evaporation direction (Figure 3.5a). This is variously termed planar, parallel or homogenous alignment. In commercial displays, 'silicon monoxide' has been used as the evaporated layer and gives a surface of very high thermal stability.

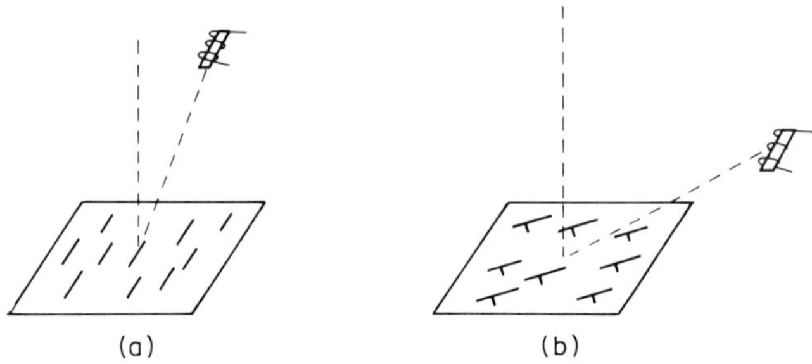

Figure 3.5 Alignment geometry from evaporated inorganic layers.

(b) Low-tilt alignment. If an evaporated planar aligning layer is mechanically rubbed with a cloth or similar material along the preferred alignment direction, a small tilt is induced in the alignment[57]. The sense is such that the director is

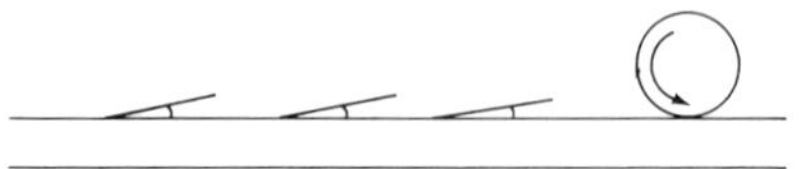

Figure 3.6 Pretilt induction on a rubbed alignment layer.

tilted up from the substrate in the direction of rubbing (Figure 3.6). An alternative method[58, 59] of obtaining a similar result is to precoat the substrate with a plastic layer; subsequent rubbing defines the alignment direction as well as inducing the pretilt angle. A rubbed glass surface[55] may also be effective, but if an abrasive[60] is used in the rubbing process no tilt is obtained. The magnitude of the pretilt using either technique is fairly small; figures quoted of 1–5° are typical. It appears possible to control the pretilt angle within these limits by variation of the rubbing pressure and the polymer. Polyimides are commonly used in commercial displays because of their good chemical and thermal stability. Water-soluble polymers such as PVA (polyvinyl alcohol) are convenient for laboratory use.

(c) High-tilt alignment. The evaporation of an inorganic layer at moderately high angle to a substrate induces a zero tilt planar alignment, as described above. If the disposition of the evaporation source and the substrate is changed so that the deposition occurs from a glancing angle[61], the alignment direction rotates to point parallel to the evaporation direction in the plane of the substrate and angled up towards the source with a tilt of *ca.* 20–30° (Figure 3.5b). The tilt angle is material dependent. Double-evaporation techniques[62] can provide pretilt angles in the range 0–45°, but are difficult to control.

(d) Homeotropic alignment. Orientation of the director perpendicular to a surface is usually termed homeotropic alignment and can be achieved in several ways. Scrupulously cleaned glass surfaces may give homeotropic alignment, as do surfaces treated with lecithin or quaternary ammonium surfactants such as HTAB[63, 64]. To obtain more stable aligning layers, silane[65] derivatives or chromium complexes[66] which bind chemically to glass surfaces have been proposed.

(e) Tilted homeotropic alignment. An alignment mode in which the LC director is almost perpendicular to the aligning surface but with a small tilt in a defined direction is obtained[58] by depositing a surfactant onto a substrate pretreated to give a tilted alignment. Alternatively, rubbing of a homeotropic aligning layer can be employed, but this may give inconsistent results, with an untilted homeotropic or planar alignment being induced.

It is probable that tilted and non-tilted planar alignments on inorganic surfaces are mediated largely by surface topography, either along scratches in rubbed surfaces or analogous grooves between columnar structures which result from evaporation steps. When a plastic layer is present, local melting and polymer chain alignment may be invoked. Surfactants are assumed to associate with the substrate through their polar groups, leaving hydrocarbon chains pointing away from the surface. Association between these chains and the LC molecules provides the alignment mechanism. No complete explanation of LC alignment is available; surfaces may give variable results dependent on the nature of the LC, its previous history, or temperature. There is, however, no shortage of data in the literature, and the field has been reviewed recently[58, 59].

3.3.2 Twisted nematic display

The twisted nematic display[67] is overwhelmingly the dominant LC device in commercial production, and is used in virtually all LC watches and calculators available at the time of writing. This dominant position derives from its proven qualities of low voltage operation, low power consumption, good contrast and relatively good multiplexing properties. The twisted nematic (TN) cell is a field effect device, i.e. its operation does not rely on the passage of any current, and

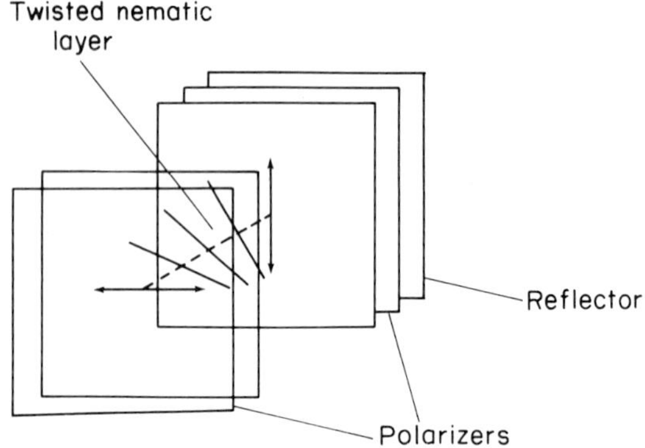

Figure 3.7 Components of a twisted nematic display cell.

this, coupled with the low operating voltage, can lead to very good operating lifetimes; it also makes the display ideal for battery-powered and portable equipment. The chief shortcomings of the TN display are common to many other LC devices: slow response at low temperature, poor brightness and reduced contrast in multiplex applications.

A TN device consists of a twisted LC layer, laminated between two polarizing films (Figure 3.7). The twist of the LC, normally close to 90°, is not obtained from a cholesteric pitch, but is defined by the alignment layers on the conducting glass plates of the cell; these are planar and mutually perpendicular. Under elastic forces, the NLC takes up a uniformly twisted configuration throughout the cell. In principle, this twist could be left- or right-handed, but the degeneracy of these states is lifted by the presence of a pretilt at the aligning surfaces and often by the addition of a small concentration of a chiral additive[57] to the NLC. The twisted birefringent structure which results has the unique optical property that light polarized parallel or perpendicular to the alignment direction on the cell wall will follow the twist of the LC director through the cell and emerge with its own plane of polarization rotated through 90° (Figure 3.8). This behaviour is discussed in more detail below. If such a twisted cell is placed between two *crossed* polarizers, the whole assembly will transmit light.

The LC used in a TN display is of positive $\Delta\varepsilon$. When a voltage is applied to the

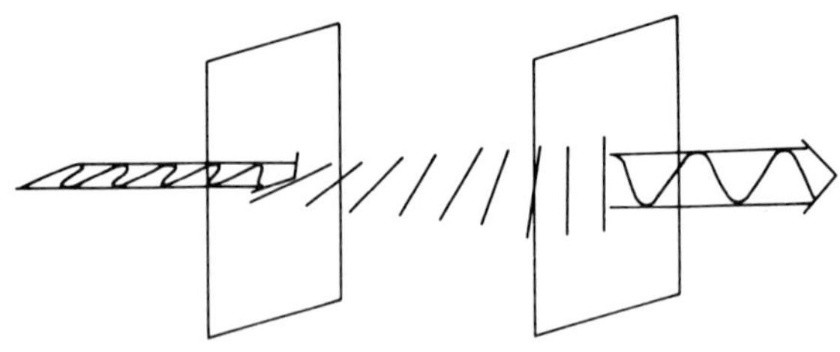

Figure 3.8 The twisted nematic cell, field 'off' state.

cell, the tendency is for the molecules near the centre of the cell to realign parallel with the field and perpendicular to the glass plates. At high voltage, this deformation spreads through the cell so that the whole LC layer is aligned parallel to the field except in a thin layer close to the aligning surfaces (Figure 3.9). In such a state, the LC no longer supports a twist and its optical activity vanishes. The cell now has little effect on the state of polarization of light, and when placed between crossed polarizers is optically extinct (Figure 3.9). Areas of a TN display which are activated by a voltage therefore appear black against a light background, and information is displayed according to patterns etched into the conductive coating on the glass plates, as described above.

Although the above account of a TN cell explains the gross features of its operation, it is too simplified to form a basis for design and selection of LC materials. Detailed experimental investigation and numerical modelling of TN devices have been carried out[68–73], and the results are in good agreement. It is therefore worth while to examine the processes involved in the switching of the TN display at some length, the results being in part applicable to other LC display modes.

The ability of a twisted LC layer to rotate the plane of polarization of light has been known since the investigations of Mauguin[55]. The simple picture of plane-polarized light being rotated by the LC is adequate when the following inequality, which is known as the 'Mauguin limit', holds:

$$u = \frac{2d\Delta n}{\lambda} >> 1 \tag{3.17}$$

where d is the cell thickness and λ the wavelength of light. For practical LC materials and device construction, however, the Mauguin limit is not attained and in this case the situation is more complex. Light 'leaks' between the two possible modes propagating in the cell and interference occurs between the emerging rays. The result is that light exits from the twist cell with an elliptical polarization[74, 75], and is imperfectly transmitted by the second polarizer. Analysis gives the equation for the transmitted light intensity, which is the basis of the 'Gooch–Tarry curve':

$$\frac{I}{I_o} = 1 - \frac{\sin^2[(\pi/2)(1 + u^2)^{1/2}]}{1 + u^2} \tag{3.18}$$

The ideal behaviour of full transmission of light (neglecting inevitable losses in the polarizers and by reflections, etc.) is obtained for specific values of u of $\sqrt{3}$, $\sqrt{15}$, $\sqrt{35}$, etc. This may be expressed in more concrete terms by taking the wavelength of peak visual sensitivity λ as equal to 0.555 μm. Optimum behaviour is then obtained for specific values of thickness and birefringence:

$$d\Delta n = 0.48, 1.07, 1.64, \text{etc.} \tag{3.19}$$

Deviation from these values brings about a reduction in the contrast of the display and, because u is a function of wavelength, may lead to undesirable colouration[69, 76] in the off state.

If we consider a TN cell without any pretilt in the surface alignment, then in the absence of an applied voltage there is equally no tilt in the director throughout the thickness of the cell. Consequently, application of a small voltage results in no net torque on the director, and no switching action is observed. This state of affairs remains until the applied voltage exceeds a critical figure, V_c (see equation 3.33), independent of the cell thickness, where a small thermally driven tilt in the centre of the cell is stabilized by dielectric forces balancing the elastic forces. Further increase of the applied voltage results[77–79] in an increase of tilt angle at the centre of the cell and a progressive tilting of nematic layers closer to the aligning surfaces. A practical TN cell with finite pretilt behaves similarly, although the onset of distortion is more gradual and strictly begins at the smallest applied voltage. Just above the critical voltage, no change is visible in the TN cell. Viewed at normal incidence, switching might typically begin at *ca.* 1.5 times the critical voltage and be essentially complete by 2 times the critical voltage. Furthermore, in a practical display, the critical voltage might be of the order 1 V, while operation of the device from a 3 V source is required. Under such circumstances, although the cell appears fully switched at normal incidence, the off-axis observations are quite different. The combination of a 90° twist and uniform pretilt in the TN cell means that the director at the centre of the cell, where the first dielectric distortion occurs, points in a defined direction at 45° to the rubbing directions on each glass plate. It is in this quadrant of the cell that the optical properties change first as voltage is applied, and at glancing incidence the optical threshold voltage is little greater than the critical voltage. In most displays it is this quadrant which is angled towards the viewer in normal use, since from other directions the display may fail to produce adequate contrast under ordinary drive conditions.

At the maximum applied voltage a TN cell still contains significant regions of the LC thickness close to the aligning surfaces where the director is far from perpendicular to the glass. In the centre of the cell, however, the molecules are sufficiently close to homeotropic alignment that the effective birefringence in this region is small, and it is energetically favourable to concentrate the twist of the LC layer into the same thickness. The cell now behaves approximately as two isolated, mutually perpendicular, birefringent slabs. Such a combination has no effect on light propagating through the cell at normal incidence, but at glancing incidence[80] the plane of polarization is rotated and the net result is that some light is passed, most seriously when the product $d\Delta n$ is large in relation to λ.

These effects conspire to limit the optical performance of the TN display. The angular variation of viewing properties is a complex function of material birefringence, cell thickness, pretilt angle, drive voltage and the dielectric and elastic parameters of the liquid crystal[70]. In practical terms, a good contrast ratio can be obtained between normal incidence and 45° to the normal in a reflective, direct drive display in the preferred viewing quadrant.

The respone time of an LC electrooptic cell on turn off has been shown[50] to be a function of both material constants and cell parameters:

$$T_{\mathrm{d}} = \frac{\eta}{k\pi^2 d^2} \tag{3.20}$$

where η is properly a rotational viscosity and k an average elastic constant. The thickness of the cell is d, and the equation holds for small deformations; however, it does also give a useful guide to response times over the TN electrooptic switching curve. The inclusion of a term in d on the denominator of the expression can be linked back to material properties by assuming a constant product $d\Delta n$ which, as discussed above, is the parameter of primary importance to the optical properties of the cell. In order to optimize the optical response time of the display, the relevant property[46] is the 'response factor' (γ) defined by

$$\gamma = \frac{\eta}{\Delta n^2} \tag{3.21}$$

which should be minimized for best results, always provided that Δn is suitable for a cell thickness which is commercially viable for production. Most cells are manufactured with $d\Delta n \simeq 1\ \mu m$, while some are produced with $d\Delta n \simeq 0.5\ \mu m$, corresponding to the first minimum of the Gooch–Tarry curve and giving a wider angle of view. Cell thicknesses of 6–12 μm are common.

(a) Multiplex drive of TN cells. Despite the complications outlined above, the directly driven TN display cell in which a separate electrical connection is made to each active segment is a forgiving technology and the constraints on the LC material component are not severe; many LC systems are now known which can provide the necessary combination of positive $\Delta\varepsilon$ and moderately high birefringence, while exhibiting an adequate temperature range and stability. Most LC displays, however, are driven by a multiplexed drive scheme, the purpose of which is to reduce the number of electrical connections made between the display and the driver circuit. This applies even in watch displays where a two-way multiplexing scheme is commonly employed. The requirements on LC properties become ever more stringent as the multiplex level is raised, and the origin and consequences of these new requirements will now be examined.

A directly driven display has one conducting glass plate patterned with the active segments of the legend, while the other plate forms a back plane which is only patterned to prevent turn-on of the cell in undesirable regions, such as over the electrical lead-ins. For a display with P active segments, $P + 1$ connections must be made to include the back plane. In complex displays this becomes costly and finally impractical as P rises, and recourse is made to multiplexing. Both front and back plates of glass are patterned, with active segments being joined together in such a way that no two segments are connected to each other on both the front and back planes. The most obvious pattern is a dot matrix of intersecting rows and columns (Figure 3.4), but seven-segment figures can be patterned in an

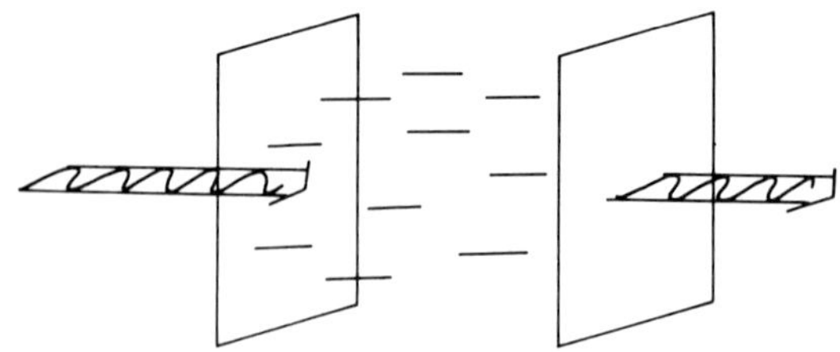

Figure 3.9 The twisted nematic cell, field 'on' state.

electrically equivalent manner. If the number of rows in the matrix is R, the necessary number of connections becomes $P/R + R$.

The realization of a multiplex driving scheme for an LC device depends on the observation[81] that the switching of LC cells is in response to the r.m.s. voltage applied, provided that the frequency of the waveform is rather faster than the response time of the cell. In order to operate a matrix-type display, it is necessary to apply waveforms to the rows and columns which allow any specified element of the matrix to be addressed with an r.m.s. voltage greater than or less than the optical threshold voltage according to whether the pixel is selected as being on or off. This addressing must be carried out irrespective of the state of other pixels in the matrix.

Considering a matrix of N rows which is multiplexed N ways, the period of the addressing waveform is first split into N equal time slots, during each of which a voltage pulse V_R is applied to one row of the matrix. All rows not receiving this strobe pulse are held at zero volts. During each time slot, each column receives a voltage pulse of $\pm V_C$ conveys information to the display. Pixels lying on rows that are not being strobed cannot discriminate the polarity of the information pulse, but if a strobe pulse is present, the voltage drop across the pixel is $V_C + V_R$ or $V_C - V_R$ according to this polarity. The state of any pixel is therefore determined by the sign of the information (column) voltage at the time when the strobe (row) voltage is applied, and not by the sign of the column voltage at any other time; all of the other $N - 1$ time slots are available and used for addressing other rows of the display (Figure 3.10). The r.m.s. voltages obtained are:

$$V_{ON} = \sqrt{\left[\frac{(V_R + V_C)^2 + (N - 1)\, V_C^2}{N}\right]} \tag{3.22}$$

$$V_{OFF} = \sqrt{\left[\frac{(V_R - V_C)^2 + (N - 1)\, V_C^2}{N}\right]} \tag{3.23}$$

It is usual to specify the relative magnitude of V_C and V_R by a select scheme; in a $1/S$ select scheme, $1/S$ of the supply voltage is applied as the information pulse and $(S - 1)/S$ as the strobe. An optimum ratio of V_{ON}/V_{OFF} is obtained[81] when the Alt–Pleshko condition is satisfied:

$$S = 1 + \sqrt{N} \tag{3.24}$$

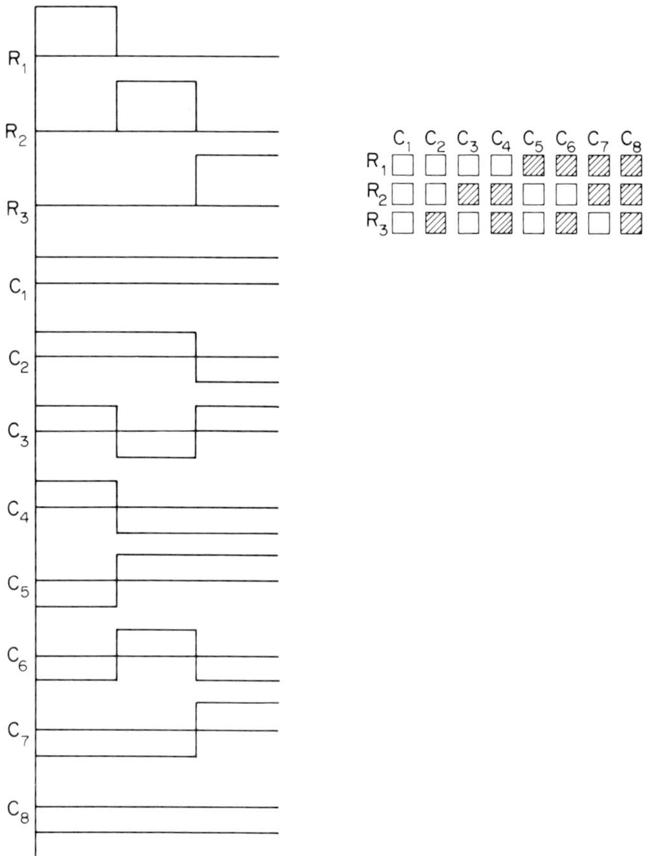

Figure 3.10 Simplified waveforms for three-way multiplex drive, allowing addressing of all possible matrix patterns.

and practical drive circuits usually choose S to be an integer close to the optimum. Using waveforms which are further optimized[82] to make the best use of supply voltage and eliminate any d.c. component, the supply voltage V_b is split between V_C and V_R without other losses, when the following expressions for the r.m.s. voltages may be obtained:

$$V_{ON} = \frac{V_b}{S} \sqrt{\left[\frac{N - 1 + S^2}{N}\right]} \tag{3.25}$$

$$V_{OFF} = \frac{V_b}{S} \sqrt{\left[\frac{N - 1 + (S - 2)^2}{N}\right]} \tag{3.26}$$

For the Alt–Pleshko scheme

$$\frac{V_{ON}}{V_{OFF}} = \sqrt{\left(\frac{\sqrt{N} + 1}{\sqrt{N} - 1}\right)} \tag{3.27}$$

Two points which emerge from the foregoing discussion should be emphasized. In a multiplexed drive scheme, a finite voltage is applied to all segments of the display including those that should be turned off. The ratio of V_{ON}/V_{OFF} is limited by the multiplex level and approaches unity as the multiplex level rises. Secondly, once the drive scheme and supply voltage are fixed, the absolute voltage levels are also determined, and the properties of the LC must be tuned to give optimum performance in each application; it remains to be considered how this can be achieved.

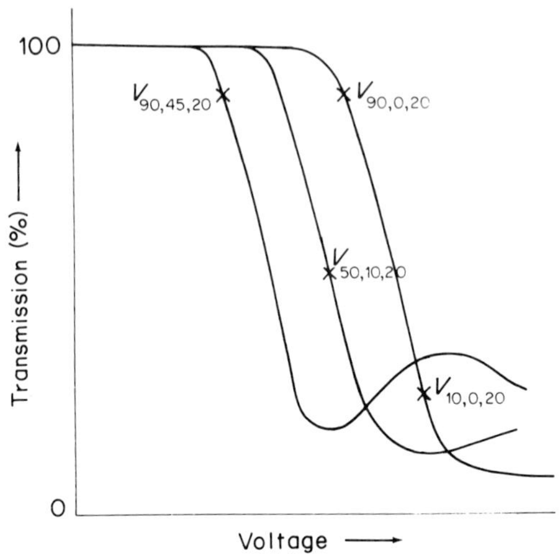

Figure 3.11 Optical transfer characteristic of a TN cell at different viewing angles.

The optical transfer characteristic of a TN cell exhibits the well-defined threshold behaviour necessary for the implementation of multiplexing (Figure 3.11), and the shape of the response curve can be characterized by the applied voltage which is necessary to obtain a defined degree of optical response. For a TN cell between crossed polarizers, such voltages are referenced by three subscripts:

$$V_{x,y,z} \tag{3.28}$$

where x is the percentage transmission of light through the cell, y is the viewing angle measured with respect to the normal axis in the lowest threshold voltage quadrant of the cell and z is the temperature in degrees Celsius. Thus $V_{90,45,20}$ is the voltage which produces 90 per cent transmission of light through the cell, measured at 20 °C and 45° off-axis, i.e. a glancing angle threshold voltage. If any lower voltage is applied to the display, its effect will not be obvious under normal viewing conditions. Similarly, it is possible to define a normal incidence saturation voltage $V_{10,0,20}$ at which the cell appears fully switched on under all normal conditions. To obtain optimum performance from a multiplexed display, the following conditions should be met:

$$V_{OFF} \leqslant V_{90,45,20} \tag{3.29}$$

and

$$V_{ON} \geqslant V_{10,0,20} \tag{3.30}$$

The situation is complicated by the fact that the threshold and saturation voltages of the cell are approximately linear functions of temperature, each reducing in magnitude as the temperature is raised. Thus to estimate the performance of a cell over a practical working range, the voltages above should be substituted by quantities such as $V_{90,45,40}$ and $V_{10,0,0}$ respectively. For a reflective TN cell, these criteria cannot both be satisfied in practice above a very low multiplex level, and are unnecessarily severe[71]. More reasonable criteria are provided by

$$\frac{V_{50,10,0}}{V_{90,45,40}} = M'_{0-40} \leqslant \frac{V_{ON}}{V_{OFF}} \tag{3.31}$$

or

$$\frac{V_{50,10,20}}{V_{90,45,20}} = M'_{20} \leqslant \frac{V_{ON}}{V_{OFF}} \tag{3.32}$$

The latter inequality is appropriate for high levels of multiplex when V_{ON}/V_{OFF} is small, and it is common practice to compensate electronically for temperature variation by changing the absolute voltage levels supplied to the display by the driving electronics. Viewed in this light, three electrooptic properties of the LC material are important for multiplexed displays, these being the threshold voltage, which must be slightly greater than V_{OFF}, the temperature dependence of the threshold, which should be small, and the threshold curve steepness, quantified by M'_{20}, which must also be small.

The threshold voltage of a TN cell mirrors the critical voltage V_c, which may be related to fundamental properties of the LC fluid:

$$V_c = \pi\left(\frac{k'}{\Delta\varepsilon\varepsilon_o}\right)^{1/2} \tag{3.33}$$

k' being an elastic constant primarily dependent on k_{11}. The threshold sharpness is related indirectly to the capacitance threshold slope, which, close to V_c, has the form

$$\text{Slope } \alpha\left(\frac{5k_{33}}{8k_{11}} + \frac{\Delta\varepsilon}{\varepsilon_\perp}\right)^{-1} \tag{3.34}$$

Detailed modelling[69–71] confirms the importance of minimizing k_{33}/k_{11}, and $\Delta\varepsilon/\varepsilon_\perp$ to optimize the available multiplex level, and also indicates the importance of the product $d\Delta n$ (referred to above) which becomes of increasing importance under multiplex drive.

(b) Materials for multiplexed TN displays. As in other areas of LC materials optimization, the challenge of devising materials for a multiplexed TN cell arises

from the desire to adjust simultaneously several physical properties of the mixture towards extreme or target values: the changes in composition required to obtain the desired parameters are often mutually exclusive or contradictory. Implicit in the selection of components for an LC mixture is the assumption that if a compound or structural class has a given physical property, it will confer that property on mixtures in which it is included. This is broadly true, although the variation of physical properties with composition in a mixture is often highly non-linear.

It is universal practice[83] in the design of materials for multiplexed TN cells to include components of both large positive and small $\Delta\varepsilon$. Adjustment of the relative proportions of these components allows the anisotropy of the resulting mixture to be finely tuned to permit alteration of the threshold voltage through equation 3.33. It is much more difficult to obtain changes in k_{11} of sufficient magnitude to achieve the same effect. Frequently, each of these two compound classes will be used in the form of low melting mixtures of homologues to minimize problems of crystallization in the final mixture. The choice of particular structural types is made on the basis of the intended use of the material. For many applications, the sharpness of switching threshold is paramount, dictating the use of materials having a minimal k_{33}/k_{11} ratio according to equation 3.34. As V_{TH} for a given application is fixed, it is difficult to make large changes in $\Delta\varepsilon/\varepsilon_{\perp}$. In other cases, a low temperature variation of threshold voltage may be most important and dictate the use of components derived from saturated ring systems, which usually possess this property. Several structural classes may, of course, be used together to obtain a compromise between properties. Finally, components will usually be required to adjust the clearing point of the mixture to a desired value.

The mixing[84] of compounds respectively high and low in $\Delta\varepsilon$ usually leads to enhancement of smectic properties, and it is common under these circumstances for combinations of purely N materials to exhibit wide S_A phase ranges (Figure 3.12). The extent and behaviour of such injected smectic phases cannot be predicted except by argument from close analogy, and empirical experimentation is the most useful approach to their control; however, they are doubly important in multiplexable mixtures. The obvious effect of an S phase is to limit the lower working temperature range of the mixture, and this may pose a very serious problem for the materials designer. What is less expected is the influence of the short range S-like ordering which appears in N phases in the vicinity of an injected S phase. This changes the absolute elastic constants[42, 85–87] so as to decrease k_{33}/k_{11} markedly and improve the attainable multiplex level; therefore it appears desirable to formulate mixtures which are on the point of becoming S but actually remain N at all temperatures. This is possible by making mixtures with a composition close to the steeply sloping side of an S hump in the phase diagram. The variation of LC physical properties in the vicinity of such an injected phase has been reported[43].

When the components of high and low $\Delta\varepsilon$ in a multiplexed mixture have been chosen, and their relative proportions fixed by the required threshold voltage, there is little possibility for large variations of the physical properties of the final

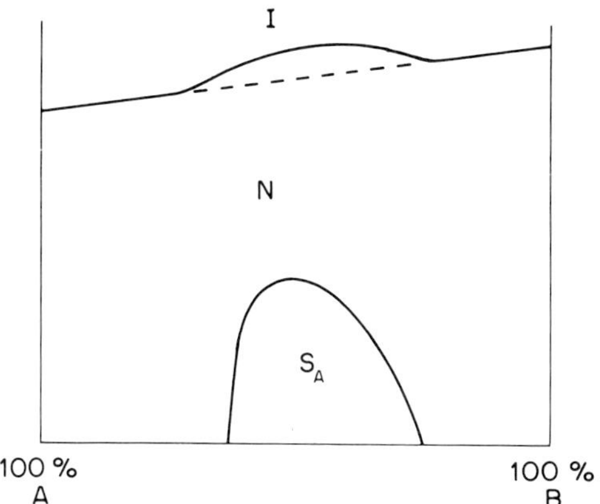

Figure 3.12 Schematic phase diagram showing injection of smectic properties into mixtures of purely nematic compounds of high and low dielectric anisotropy.

product. Furthermore, because the analysis of structure-property correlations in LCs can often be extended to structural subunits, it is frequently the case that if one property is optimized other parameters become virtually fixed. As a conspicuous example, the materials showing the lowest ratios of k_{33}/k_{11} are based on aromatic phenyl or pyrimidinyl ring systems[46, 48, 88–90], which also confer high Δn on the mixture (Table 3.2). Similarly, most materials of very high $\Delta\varepsilon$, necessary for low voltage multiplex or static drive, are based on aromatic rings, although this difficulty has been alleviated by the introduction of the 2,5-disubstituted 1,3-dioxanyl group as an LC building block. It seems established that large changes in the properties of such mixtures are dependent on the synthesis of new structural classes of LCs, while the incorporation of known compounds in mixtures in new combinations is based largely on empirical experiment and can produce incremental improvements in the final mixture.

(c) Materials for wide temperature range TN displays. The method of formulation of LC mixtures described above is a general one[91]; components are selected according to their physical properties as pure compounds or their behaviour in mixtures determined by empirical experiment. Optimization of the mixture is carried out by varying the relative concentrations of the components and attempting to make use of non-linearities in the variation of different properties with composition, this process also being guided largely by experience gained from empirical experiment. The technique has been applied to LC mixtures for application over wide temperature ranges, a typical target being operation between −30 and +85 °C.

The upper limit to the working range of an LC mixture is the clearing point, or the drastic pretransitional fall in Δn which occurs close to this temperature; extension of the working range upwards is therefore easy, at least in concept, by making use of several high clearing point components according to equation 3.1.

Assuming that crystallization and S phases do not intervene, the lower limit of the working range is determined by the rise in viscosity as the temperature is lowered, which dominates any change in elastic constants to increase the response time of the display (equation 3.20) beyond what is reasonably acceptable. The minimization of the viscosity of the mixture and its variations with temperature are therefore primary considerations, although elastic constant variations may contribute a 100 per cent variation in response times.

The fundamental approach to this problem is the separation of the mixture into a component of positive $\Delta\varepsilon$ and a low viscosity material, it being difficult to combine low viscosity with a highly dipolar nature in a molecule large enough to display LC properties. A further component is usually needed to raise the clearing point of the mixture, and again a non-polar additive is usually chosen to minimize viscosity (see Table 3.3). The choice of components, then, is largely dictated by a compromise between clearing point and viscosity at low temperatures, but other factors such as birefringence and volatility play a part. It has been shown[92]

Table 3.3 Components for use in wide working range TN mixtures

(a) Low viscosity components

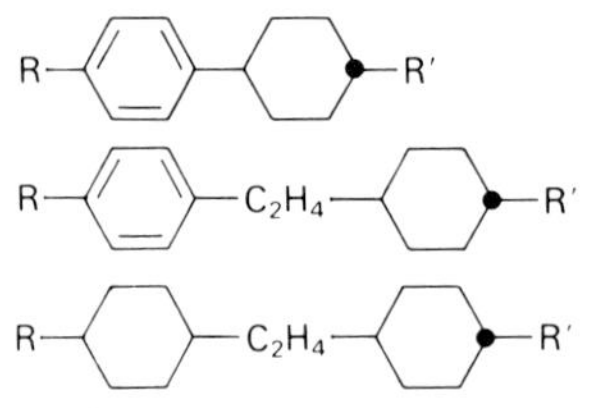

R—Ph—Cy—R′

R—Ph—C_2H_4—Cy—R′

R—Cy—C_2H_4—Cy—R′

(b) High clearing point components

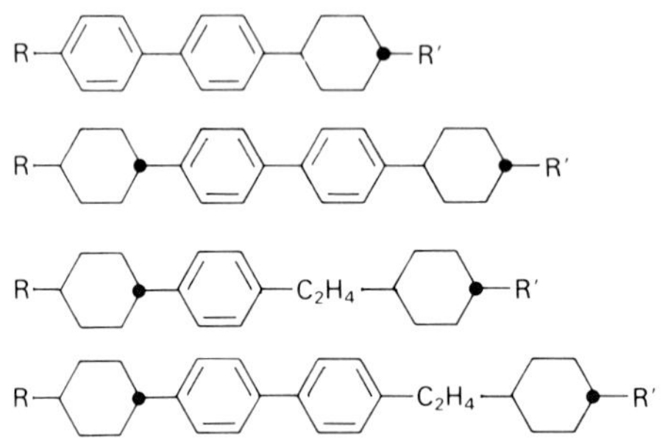

R—Ph—Ph—Cy—R′

R—Cy—Ph—Ph—Cy—R′

R—Cy—Ph—C_2H_4—Cy—R′

R—Cy—Ph—Ph—C_2H_4—Cy—R′

(c) Wide range, low viscosity nematogens

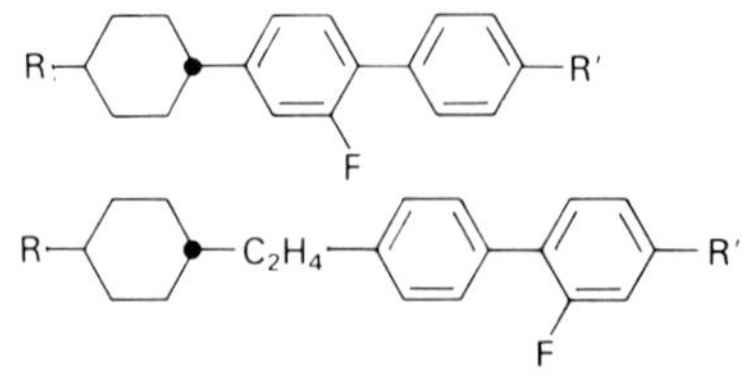

R—Cy—Ph(F)—Ph—R′

R—Cy—C_2H_4—Ph—Ph(F)—R′

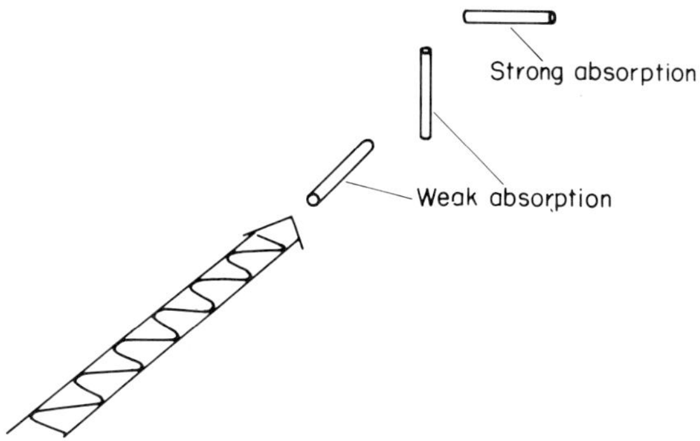

Figure 3.13 Selective absorption of polarized light by an aligned dichroic dye.

that the properties of low viscosity and high clearing materials can be combined into a single structural class, so partly resolving such difficulties.

3.3.3 Heilmeier GH display

It has already been pointed out that many dyes in solution in NLC take on the ordering of the host and show dichroism (Figure 3.13). Such a dyed LC layer may be placed[93, 94] in an electrooptic cell similar to that used for the direct drive TN display, provided with a single polarizer with its easy axis aligned with the rubbing direction on the adjacent glass plate (Figure 3.14). This assembly will absorb light efficiently around the λ_{max} of the dye and will appear deeply coloured until a voltage applied to the cell switches the LC dye into a homeotropic alignment. The dye no longer absorbs efficiently in this orientation and selected segments stand out as white areas on a coloured ground. The 90° twist in the cell is not necessary for this display effect to operate, although it improves the steepness of the electrooptic threshold curve by optically isolating the boundary layers in the

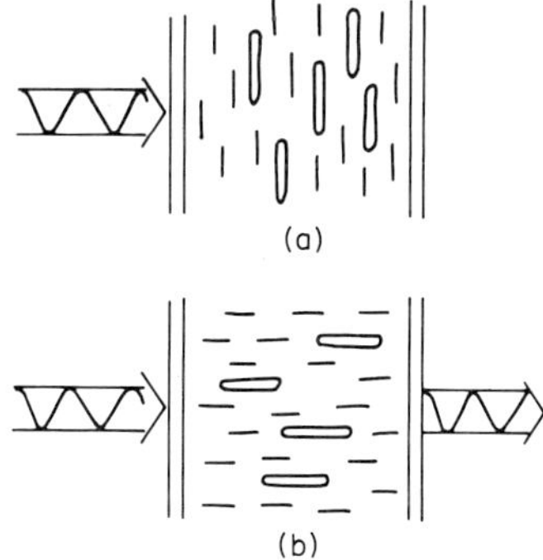

Figure 3.14 Heilmeier guest–host cell in (a) 'off' and (b) 'on' states.

part-switched cell[95], but the Heilmeier display does not have a steep electrooptic curve and is poorly suited to multiplex drive. A similar effect with reversed contrast is available using tilted homeotropic alignment and an LC of negative $\Delta\varepsilon$; the director is then switched into a planar alignment by application of a voltage and active elements appear as coloured areas on a light background. Primary requirements for the materials used in a Heilmeier cell are similar to those for a static drive TN display — relatively large (positive or negative) $\Delta\varepsilon$ and a moderate viscosity. Birefringence is a less serious consideration, but the added dimension of the dichroic dyestuff must be considered. The dye must exhibit adequate solubility and extinction coefficient to provide a high degree of light absorption in the cell. The optical order parameter is also crucial, being related to the available contrast ratio:

$$\mathrm{CR} = \frac{A_{\parallel}}{A_{\perp}} = \frac{1 + 2S}{1 - S} \tag{3.35}$$

In guest–host systems it is therefore common practice to raise the clearing point of the host well above the limit of the operating temperature range of the display in order to maximize S.

Stable pleochroic dyes are now available with adequate order parameter in a range of colours[26–28], but it is frequently required to match the dye used to a specified standard colour or obtain a neutral (black and white) appearance. In order to achieve these targets, mixtures of dyes must be used. Given the absorption spectrum of each dye, it is possible to relate the perceived colour and lightness of a mixture in a display cell to its composition and concentration by reference to the CIE 1931 standard observer and chromaticity diagram[96], which together with their successors represent an extension and quantification of the familiar colour-mixing triangle. The perceived colour change as a function of illumination is also described by the CIE standards, and criteria for obtaining neutral coloured dye displays under varying illumination conditions have been presented[97].

3.3.4 White–Taylor GH display

The Heilmeier display gives the opportunity for obtaining colour in an LC device and dispenses with one of the polarizers necessary on a TN cell. However, the remaining polarizer limits the brightness of the display, and the White–Taylor display provides a means to overcome this, together with an improved (almost uniform) angle of view characteristic[98].

The display uses a cholesteric to nematic phase change induced by an electric field in a dyed LC layer. The LC used is a guest–host system similar to that used for the Heilmeier display, but doped with an optically active additive to give a Ch phase with a pitch in the region of micrometres. Under the influence of aligning layers similar to those used in the TN cell, the LC adopts a configuration with the helical axis perpendicular to the glass walls of the cell giving a uniform planar texture. If the pitch of the LC helix is relatively short ($P\Delta n \ll 1$), the light entering the cell is not guided as in the TN display, but passes through the LC, being strongly absorbed in those regions where the local director orientation (and dye transition moment) is suitably aligned. The process is formally described by considering the propagation of two elliptically polarized modes, allowing the prediction[98–100] of the behaviour of the cell in the important region where light is substantially, but imperfectly, guided by the LC. The qualitative result is that efficient absorption of light requires either a short helical pitch or an LC of very low birefringence. A dye concentration higher than that used in the Heilmeier cell is necessary, but the polarizer is redundant.

When a voltage is applied to the cell, there is first a change in the orientation of the helical axes under dielectric forces and the helix then progressively unwinds to a homeotropic N texture similar to the 'on' state of the Heilmeier cell. This process is essentially complete at the critical voltage,

$$V_c = \frac{\pi^2 d}{P}\left(\frac{k_{22}}{\Delta\varepsilon\varepsilon_o}\right)^{1/2} \qquad (3.36)$$

although a higher voltage may still be necessary to reorientate the LC close to the aligning surfaces and generate maximum contrast. Because the dye concentration required in the White–Taylor cell is high and the order parameter of available pleochroic dyes is lower than that of a polarizer, the 'on' state is more highly coloured than in a Heilmeier display and comparison of the performance of the two devices depends on assessment of the relative importance of contrast and brightness to the user.

There are important variations possible to the geometry described above. An essentially similar display is obtained using homeotropic boundary conditions in the cell, the liquid crystal director being perpendicular to the glass with the pitch unwound at the surfaces, but undergoing a distortion to give a helical structure whose axis is perpendicular to the glass near the centre of the cell. Also of interest[101] is the homeotropic phase change cell utilizing a very long pitch dyed Ch LC with negative $\Delta\varepsilon$. In the absence of any applied voltage the helix of such an LC can be fully unwound by the boundary forces provided that the following inequality holds:

$$\frac{k_{33}}{2k_{22}} > \frac{d}{P} \qquad (3.37)$$

When the director tilts under the influence of an applied field, the helix winds up to approach its natural pitch and the structure absorbs light in the same way as the 'off' state of the White–Taylor display, giving a positive contrast display. The requirement of equation 3.37, given that typically $2k_{22} \simeq k_{33}$, means that $P \simeq d$ and only *ca.* one helical turn can be present in the 'on' state of the display.

The requirements of the White–Taylor display for dichroic materials with a large magnitude of $\Delta\varepsilon$ are similar to those demanded by the Heilmeier cell, but dyed phase change displays present some unique targets also. The induction of a

Table 3.4 Chiral additives for phase change display use

	Pitch (μm)
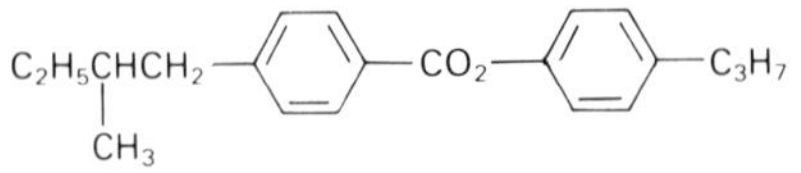 $C_2H_5CHCH_2$ (CH_3) — CO_2 — C_3H_7	0.24
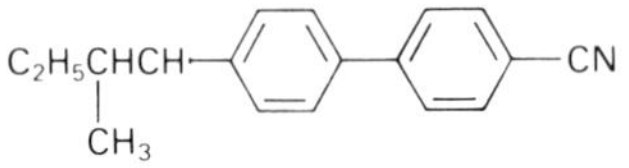 C_2H_5CHCH (CH_3) — CN	0.14
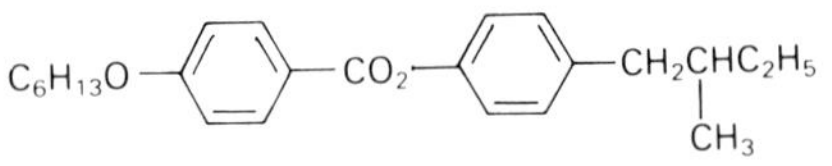 $C_6H_{13}O$ — CO_2 — $CH_2CHC_2H_5$ (CH_3)	0.23
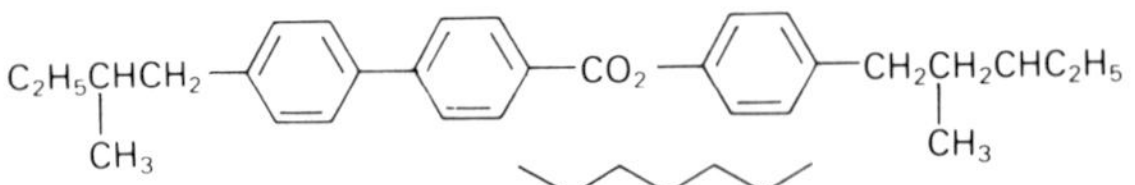 $C_2H_5CHCH_2$ (CH_3) — CO_2 — $CH_2CH_2CHC_2H_5$ (CH_3)	0.9
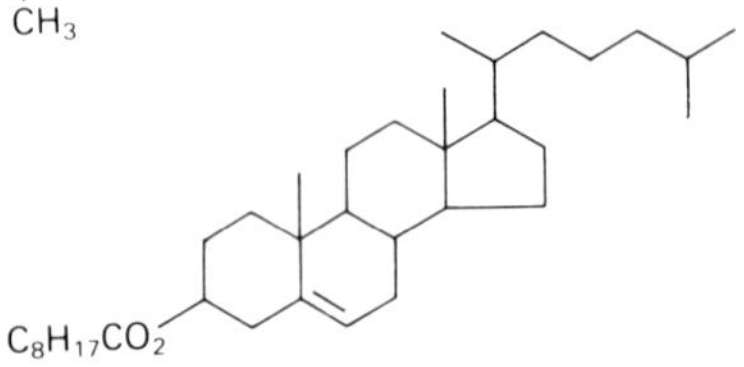 $C_8H_{17}CO_2$	0.16

helical pitch in the mixture requires the addition of a chiral component, but several compounds are available which give a suitable pitch at concentrations of $\lesssim$ 10 per cent, so that other physical properties of the mixture are not too seriously changed (see Table 3.4). In order that efficient absorption of light can be obtained, a low Δn is necessary, and in the single-turn positive contrast device, this requirement becomes paramount. In practical terms, this interaction of contrast and operating voltage with helical pitch, Δn, dye order parameters and concentration, cell thickness and $\Delta\varepsilon$ makes the optimization of fluids for dyed phase change displays a complex target.

3.3.5 Liquid crystal scattering displays

The anisotropy of refractive index in liquids makes possible a strong scattering of incident light without the necessity for any dye or polarizing film, and at least three separate display modes are available that make use of the phenomenon. The desirable LC properties for each are different, apart from a common requirement for fluids of very high Δn to maximize contrast, combined with reasonably low viscosity to minimize response times.

The dynamic scattering display was the most widespread LC electrooptic device prior to the introduction of the TN effect, and alone of the display effects discussed in this chapter it relies on the passage of a current through the LC. Dynamic scattering refers[102] to the turbulent state obtained in an NLC of negative $\Delta\varepsilon$ which has been doped to a resistivity of $\lesssim 10^{10}\ \Omega$ cm on application of a voltage. No polarizer is required and the brightness of the display is very good, but its legibility is heavily dependent on viewing conditions. A range of optical effects[15] deriving from the interaction of dielectric and conductivity forces can also be observed in these cells. An LC of modest viscosity, negative $\Delta\varepsilon$ and high Δn is required which can act as a good solvent for an ionic conductivity dopant. Dynamic scattering displays are now of largely historical interest as it is difficult for them to compete with TN technology in terms of power consumption and reliability.

A commonly observed defect in the White–Taylor phase change display can be exploited to give a scattering device. When a phase change cell relaxes from the field 'on' state, the Ch helices are often formed in random orientations, forming a scattering texture which can also be obtained by applying a low voltage to the field 'off' state. Switching is carried out between this scattering texture and the clear field 'on' state, multiplexing being possible[103] using the limited memory of the clear and scattering states. Material requirements are similar to those for the host in a White–Taylor device, apart from the need for a high Δn.

A recently announced[104, 111] light-scattering electrooptic effect called the nematic curvilinear aligned phase (NCAP) mode is observed in plastic sheets in which an NLC of strong positive $\Delta\varepsilon$ is dispersed as small (*ca.* 1–10 μm) droplets. The LC has a large Δn and n_o is matched to the refractive index of the polymer matrix. Light propagating through the sheet is strongly scattered by the randomly aligned NLC in the droplets, until a voltage is applied between transparent electrodes laminated onto the plastic, when the N phase aligns with its director parallel to the field and perpendicular to the electric vector of the light traversing the sheet (Figure 3.15). There is now no effective difference in refractive index between the LC and its supporting matrix, and the sheet appears clear. The necessary physical properties of the LC are as stated above, the principal difficulty being in combining a large Δn with an n_o sufficiently small to index-match common polymers that are compatible with the fabrication process. A dichroic

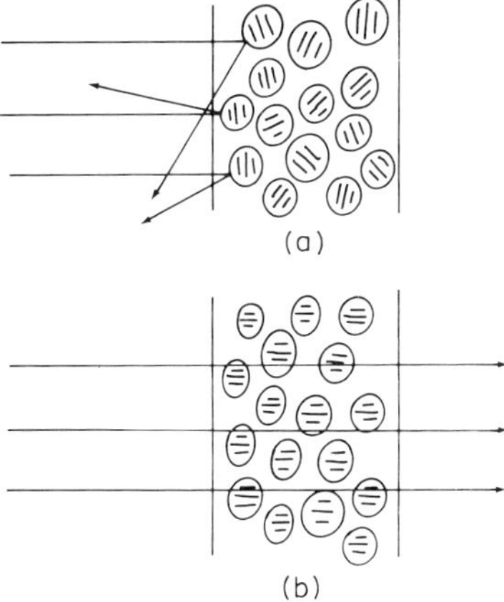

Figure 3.15 (a) 'Off' and (b) 'on' states of the NCAP light-scattering display.

dyed form of the display requiring low Δn and the same index-matching criterion is also possible. It is envisaged that the major use of the device will be in large-area outdoor applications.

3.3.6 Variable birefringence modes in LC cells

The transition[105] between the homeotropically aligned and planar states of an LC cell can be driven in either direction by an applied voltage according to the $\Delta\varepsilon$ of the fluid used, and is accompanied by a change in the effective birefringence of the cell, Δn_{eff}, from zero to a maximum value of $d\Delta n$. Placed between crossed polarizers with the planar alignment axis offset at 45° to each, the transmission of light is given by the equation:

$$\frac{I}{I_o} = \sin^2\left(\frac{2\pi\Delta n_{\text{eff}}}{\lambda}\right)$$

and because $d\Delta n$ can be several times λ for practical LC materials and display cells, a slow increase of the voltage applied to the device leads[106] to a series of wavelength-dependent maxima and minima in the transmitted light intensity. In display applications, it is usual to utilize a negative $\Delta\varepsilon$ LC, initially homeotropically aligned, so that the zero voltage birefringence is zero (Figure 3.16); the device is termed a DAP (distortion of aligned phase) cell. The electrooptic threshold curve towards the first transmission maximum can be very steep, as it occurs over a small proportion of the voltage range needed to realign the LC fully; multiplexing at a high level is possible, but the appearance of the display is sensitive to lighting conditions, temperature (through the effect on Δn) and viewing angle. The threshold voltage is determined by the same equation as V_c in the TN display (equation 3.33), taking $\Delta\varepsilon$ as negative, and the response time

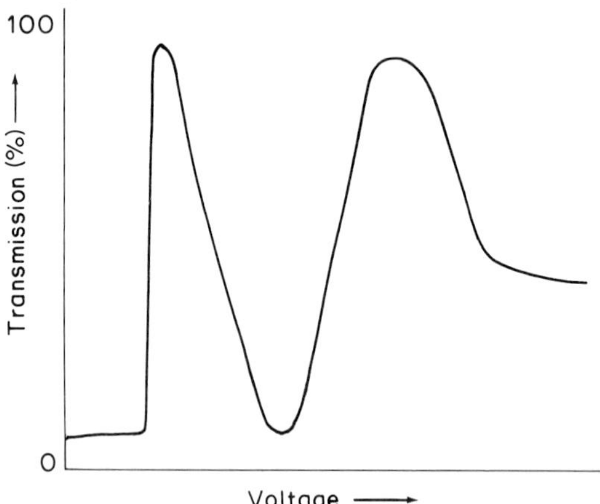

Figure 3.16 Optical response of the DAP nematic cell.

is also analogous (equation 3.20), but the inversion of the distortion geometry means that $k' = k_{33}$ and that k_{33}/k_{11} must be *maximized* to optimize the threshold slope. In practical terms, it is a compromise between viscosity, Δn and $\Delta\varepsilon$ that determines the selection of materials for these displays.

An elegant variable birefringent cell[107] based on the switching of surface layers in an N of positive $\Delta\varepsilon$ has been developed recently. By imposing a 180° bend on the director, a considerable improvement both in viewing angle and response speed can be achieved. The device is used for colour switching applications, but is not so suitable for multiplexed devices. Materials are selected on the basis of available response speed, the choice being weighed by the desire to produce an optically thin cell to minimize uniformity constraints; materials of very low viscosity and moderate Δn are therefore preferred.

An important variable birefringence display is provided by the SBE[108] (supertwist birefringent effect) cell, which is replacing TN displays in high level multiplex applications. A cell similar to that used for the TN display is used in conjuction with a long pitch cholesteric material of positive $\Delta\varepsilon$ to give a twist angle of *ca* 180–270°. The effect of increasing the twist angle in this way is to increase the slope of the switching threshold curve; at a critical twist angle dependent on material and cell parameters, this slope becomes infinite and the display is capable of being multiplex driven at very high levels, while at still larger twist angles bistable behaviour is observed[110]. A practical maximum value of the twist angle which can be used is set by the appearance of a two-dimensional scattering texture in cells. This is most pronounced in displays with large twist angles and small surface tilt angles, and the difficulty of obtaining a high surface tilt which is stable and reproducible in production, without recourse to expensive evaporative techniques has so far limited most commercially produced displays to twist angles below *ca*. 240°. Such displays are often denoted STN (supertwisted nematic) mode. Like other birefringent displays, the optical properties of these devices are highly wavelength dependent, resulting in coloured ON and OFF states.

The SBE display is a complex device, and understanding of its operation has been greatly aided by numerical modelling of the director response and optics.[113] Analytical expressions have now been derived which give a good guide to the threshold voltage[114] and off-state transmission.[115] To achieve optimum performance, all three elastic constants as well as the dielectric anisotropy and birefringence must be considered, but the requisite values depend strongly on the twist angle in the display. For twist angles around 270°, the slightly bistable behaviour which optimises the attainable multiplex level can readily be achieved using common LC materials; at smaller twist angles this becomes increasingly difficult, requiring unreasonably large values of $k_{33}/k_{\parallel}$ and k_{33}/k_{22}. Fortunately a reasonable contrast can still be obtained at high multiplex level when bistability is lost,[117] and material selection becomes dependent on the resulting threshold steepness and LC viscosity and birefringence. Variations on the SBE display have been proposed, including a dichroic dyed display,[109,119] and a birefringent display using a low $d\Delta n$ product[118] which is claimed to give better colour properties.

Alone of the display modes discussed above the SBE device is now displacing TN displays from an established application range, and these displays together account for the great majority of commercial LC device manufacture. Dichroic displays for signboard and aerospace use are made on a smaller scale. In the future, the construction of LC devices on a semiconductor backplane will remove the problems of addressing complex panels and full colour displays are available using this technology with TN cells.

Any new display mode introduced will have to offer clear benefits to compete with the established devices; the large area capability of the NCAP display is an example. Such a new display must also be capable of large scale manufacture, and utilise a LC with physically realisable properties. Until a new technology is introduced, the desire of display users to extend the application of LC displays will ensure the continued improvement of materials and devices for some years to come.

3.4 References

1. G.H. Heilmeier, L.A. Zanoni and L.A. Barton, *Appl. Phys. Lett.*, **13**, 46 (1968).
2. J.J. Wysocki, J. Adams and W. Hass, *Phys. Rev. Lett.*, **20**, 1024 (1968).
3. M. Schadt and W. Helfrich, *Appl. Phys. Lett.*, **18**, 127 (1971).
4. W. Maier and K. Markau, *Z. Phys. Chem. (Frankfurt am Main)*, **28**, 190 (1961).
5. H. Kelker and B. Scheurle, *J. Phys. (Paris)*, **4**, 104 (1969).
6. H. Kelker, B. Scheurle, R. Hatz and W. Bartsch *Angew. Chem., Int. Ed. Engl.*, **9**, 962 (1970).
7. G.W. Gray, K.J. Harrison and J.A. Nash, *Electron. Lett.*, **9**, 130 (1973).
8. G.W. Gray, *J. Phys. (Paris)*, **36**, 337 (1975).
9. C. Hilsum in *Technology of Chemicals and Materials for Electronics*, ed. by E.R. Howells, Ellis Horwood, Chichester, England (1984), Ch. 3.
10. E.C. Hsu and J.F. Johnson, *Mol. Cryst. Liq. Cryst.*, **21**, 273 (1973).
11. D. Demus, C.H. Fietksu, E. Schubert and H. Keheen, *Mol. Cryst. Liq. Cryst.*, **25**, 215 (1974).
12. D.S. Hulme and E.P. Raynes, *J. Chem. Soc., Chem. Commun.*, **1974a**, 98.
13. A.V. Ivashchenko, V.V. Titov and E.I. Kovshev, *Mol. Cryst. Liq. Cryst.*, **33**, 195 (1976).
14. J. Szulc, Z. Witkiewicz and R. Dabrowski, *Mol. Cryst. Liq. Cryst.*, **109**, 125 (1984).
15. P.G. de Gennes, *The Physics of Liquid Crystals*, Clarendon Press, Oxford (1974).
16. W.H. de Jeu, *Physical Properties of Liquid Crystalline Materials*, Gordon and Breach, New York (1980).
17. W. Maier and A. Saupe, *Z. Naturforsch.*, **14A**, 882 (1959).
18. W. Maier and A. Saupe, *Z. Naturforsch.*, **15A**, 287 (1960).
19. G.R. Luckhurst and C. Zannoni, *Nature (London)*, **267**, 412 (1977).
20. A. Buker and W.H. de Jeu, *J. Phys. (Paris)*, **43**, 361 (1982).
21. M.A. Cotter, *Mol. Cryst. Liq. Cryst.*, **97**, 29 (1983).
22. I. Haller, *Prog. Solid State Chem.*, **10**, 103 (1975).
23. K.J.A. Tough and M. Bradshaw, *J. Phys. (Paris)*, **44**, 447 (1983).
24. J.W. Emsley (ed.), *NMR of Liquid Crystals*, Reidel, Darmstadt (1985).
25. F.C. Saunders, L. Wright and M.G. Clark in *Liquid Crystals and Ordered Fluids*, Vol. 4, ed. by J.F. Johnson and A.D. Griffin, Plenum Press, New York (1984), p. 831.
26. J. Cognard, T. Hieu Phan and N. Basturk, *Mol. Cryst. Liq. Cryst.*, **91**, 327 (1983).
27. R. Kiefer and G. Baur, *Eurodisplay Proc. (Paris)* (1984).
28. F.C. Saunders, K.J. Harrison, E.P. Raynes and D.J. Thompson, *Inst. Elect. Electron. Engrs., Trans. Electron Devices*, **30**, 499 (1983).
29. R. Cox, *Mol. Cryst. Liq. Cryst.*, **55**, 1 (1979).
30. M.G. Pellatt, I.H.C. Roe and J. Constant, *Mol. Cryst. Liq. Cryst.*, **59**, 299 (1980).

31. F. Jones and T. Reeve, *J. Soc. Dyers Colour*, **1979**, 352.
32. S. Aftergut and H. Cole, *Mol. Cryst. Liq. Cryst.*, **78**, 271 (1981).
33. N. Basturk, J. Cognard and T. Hieu Phan, *Mol. Cryst. Liq. Cryst.*, **95**, 71 (1983).
34. W. Maier and G. Meier, *Z. Naturforsch.*, **16A**, 262 (1961).
35. W.H. de Jeu, *Philos. Trans R. Soc. London, A*, **309**, 217 (1983).
36. K. Toriyama and D. Dunmur. *Mol. Phys.*, **56**, 479 (1985).
37. J. Dubois and A. Beguin, *Mol. Cryst. Liq. Cryst.*, **47**, 193 (1978).
38. T. Inukai, K. Furukawa, H. Inoue and K. Terashima, *Mol. Cryst. Liq. Cryst.*, **94**, 100 (1983).
39. M. Osman and T. Hyunh Ba, *Mol. Cryst. Liq. Cryst.*, **82**, 203 (1983).
40. M. Schadt, *Mol. Cryst. Liq. Cryst.*, **89**, 77 (1982).
41. R. Eidenschinck, G. Hass, M. Roemer and B. Scheuble, *Angew. Chem., Int. Ed. Engl.*, **23**, 147 (1984).
42. H. Schad and M. Osman, *J. Chem. Phys.*, **79**, 5710 (1983).
43. D. Dunmur, R. Walker and P. Palffy-Muhoray, *Mol. Cryst. Liq. Cryst.*, **122**, 321 (1985).
44. J. Grupp, *Phys. Lett.*, **92A**, 373 (1983).
45. F. Leenhouts, H. Roebers, A. Dekker and J. Jonker, *J. Phys. (Paris)*, **40**, 291 (1979).
46. M. Bradshaw, D.G. McDonnell and E.P. Raynes, *Mol. Cryst. Liq. Cryst.*, **70**, 289 (1981).
47. M. Osman, M. Schad and H. Zeller, *J. Chem. Phys.*, **78**, 906 (1983).
48. M. Bradshaw and E.P. Raynes, *Mol. Cryst. Liq. Cryst.*, **91**, 145 (1983).
49. N.H. Hartshorne and A. Stuart, *Crystals and the Polarising Microscope*, 4th ed., E. Arnold, London (1970).
50. E. Jakeman and E.P. Raynes, *Phys. Lett.*, **A30**, 69 (1972).
51. M. Schadt and F. Muller, *Inst. Elect. Electron. Engrs., Trans. Electron. Devices*, **25**, 1125 (1978).
52. M.G. Clark, *Mol. Cryst. Liq. Cryst.*, **127**, 1 (1985).
53. J. Constant and E.P. Raynes, *Mol. Cryst. Liq. Cryst.*, **62**, 115 (1980).
54. B. Bahadur, *Mol. Cryst. Liq. Cryst.*, **109**, 1 (1984).
55. C. Mauguin, *Bull. Soc. Franç Minéral. Cryst.*, **34**, 71 (1911).
56. J. Jannini, *Appl. Phys. Lett.*, **21**, 173 (1972).
57. E.P. Raynes, *Electron. Lett.*, **10**, 141 (1974).
58. J. Cognard, *Mol. Cryst. Liq. Cryst.*, **78** (Suppl. 1), 1 (1981).
59. J. Castellano, *Mol. Cryst. Liq. Cryst.*, **94**, 33 (1983).
60. D. Berreman, *Mol. Cryst. Liq. Cryst.*, **23**, 215 (1973).
61. E. Guyon, P. Pieranski and M. Boix, *Appl. Phys. Lett.*, **25**, 479 (1974).
62. E.P. Raynes, D. Rowel and I. Shanks, *Mol. Cryst. Liq. Cryst.*, **34**, 105 (1976).
63. F.J. Kahn, US Pat. 3,694,053 (1972).
64. K. Hiltrop and H. Stegemeyer, *Mol. Cryst. Liq. Cryst.*, **49**, 61 (1978).
65. F.J. Kahn, *Appl. Phys. Lett.*, **22**, 386 (1973).
66. S. Matsumoto, M. Kawamoto and N. Kaneko, *Appl. Phys. Lett.*, **27**, 268 (1975).
67. M. Schadt and W. Helfrich, *Appl. Phys. Lett.*, **18**, 127 (1971).
68. C.V. Van Doorn, C. Gerritano and J. de Klerk in *The Physics and Chemistry of Liquid Crystal Devices*, ed. by G. Sprokel, Plenum Press, New York (1980).
69. F.J. Kahn and H. Birecki Gerritsma in *The Physics and Chemistry of Liquid Crystal Devices*, ed. by G. Sprokel, Plenum Press, New York (1980).
70. G. Baur, *Mol. Cryst. Liq. Cryst.*, **63**, 45 (1981).
71. B. Needham, *Philos. Trans. R. Soc. London, A*, **309**, 179 (1983).
72. S. Shikata, M. Isogai, K. Iwasaki and A. Mukoh, *Mol. Cryst. Liq. Cryst.*, **108**, 339 (1984).
73. G. Baur, *Eurodisplay Proc. (Paris)*, (1984).
74. C. Gooch and H. Tarry, *Electron. Lett.*, **10**, 2 (1974).
75. C. Gooch and H. Tarry, *J. Phys. D., Appl. Phys.*, **8**, 1575 (1975).
76. A. Miyaji, H. Mada and S. Kobayashi, *Mol. Cryst. Liq. Cryst.*, **74**, 121 (1981).
77. H. Deuling, *Mol. Cryst. Liq. Cryst.*, **19**, 123 (1972).
78. H. Gruler, T. Scheffer and G. Meier, *Z. Naturforsch.*, **27A**, 966 (1972).
79. H. Deuling, *Mol. Cryst. Liq. Cryst.*, **27**, 81 (1975).
80. D. Berreman, *J. Opt. Soc. Am.*, **63**, 1374 (1973).
81. P. Alt and P. Pleshko, *Inst. Elect. Electron. Engrs., Trans. Electron. Devices*, **21**, 146 (1974).
82. H. Kawakami, Y. Nagae and E. Kaneko, *Proc. Biennial Display Conf.* (1976).
83. K. Toriyama, K. Suzuki, T. Nakagomi, T. Ishibashi and K. Odawara in *The Physics and Chemistry of Liquid Crystal Devices*, ed. by G. Sprokel, Plenum Press, New York (1976).

84. C. Oh, *Mol. Cryst. Liq. Cryst.*, **42**, 1 (1977).
85. B. Scheuble, G. Baur and G. Meier, *Mol. Cryst. Liq. Cryst.*, **68**, 57 (1981).
86. M. Bradshaw and E.P. Raynes, *Mol. Cryst. Liq. Cryst.*, **99**, 107 (1983).
87. M. Bradshaw and E.P. Raynes, *J. Phys. (Paris)*, **45**, 157 (1984).
88. M. Bradshaw and E.P. Raynes, *Mol. Cryst. Liq. Cryst.*, **72**, 35 (1981).
89. H. Schad and M. Osman, *J. Chem. Phys.*, **75**, 880 (1981).
90. M. Schadt and P. Gerber, *Z. Naturforsch.*, **37A**, 165 (1982).
91. J.D. Margerum, A.M. Lackner, J.E. Jensen, L.J. Miller, W.H. Smith Jr, S.-M. Wong and C.I. Van Ast, *Mol. Cryst. Liq. Cryst.*, **111**, 103 (1984).
92. P. Balkwill, D. Bishop, A. Pearson and I. Sage, *Mol. Cryst. Liq. Cryst.*, **123**, 1 (1985).
93. G. Heilmeier and L. Zanoni, *Appl. Phys. Lett.*, **13**, 91 (1968).
94. G. Heilmeier, J. Castellano and L. Zanoni, *Mol. Cryst. Liq. Cryst.*, **8**, 293 (1969).
95. T. Scheffer, *Philos. Trans. R. Soc. London, A*, **309**, 189 (1983).
96. D. Judd and G. Wyszecki, *Colour in Business, Science and Industry*, 3rd ed., John Wiley, New York (1975).
97. T. Scheffer, *J. Appl. Phys.*, **53**, 257 (1982).
98. D. White and G. Taylor, *J. Appl. Phys.*, **45**, 4718 (1974).
99. H. Cole and S. Aftergut, *J. Chem. Phys.*, **68**, 896 (1978).
100. T. Scheffer and J. Nehring in *The Physics and Chemistry of Liquid Crystal Devices*, ed. by G. Sprokel, Plenum Press, New York (1981).
101. F. Gharadjedaghi, *Mol. Cryst. Liq. Cryst.*, **68**, 127 (1981).
102. G. Heilmeier, L. Zanoni and L. Barton, *Appl. Phys. Lett.*, **13**, 46 (1968).
103. W. Blackburn, *J. Phys. D, Appl. Phys.*, **13**, 1785 (1980).
104. J. Fergason, *Proc. SID Conf. Orlando* (1985).
105. V. Fréedericks and V. Zolina, *Trans. Faraday Soc.*, **29**, 919 (1933).
106. T. Scheffer in *Nonemissive Electro-optic Displays*, ed. by A. Kmetz and F. Von Willisen, Plenum Press, New York (1975).
107. P. Bos and K. Koehler-Beran, *Mol. Cryst. Liq. Cryst.*, **133**, 329 (1984).
108. T. Scheffer and J. Nehring, *Appl. Phys. Lett.*, **45**, 1021 (1984).
109. C. Waters, V. Brimmell and E.P. Raynes, *SID Symp. Dig.*, **25**, 261 (1984).
110. R. Thurston, *Mol. Cryst. Liq. Cryst.*, **122**, 1 (1985).
111. P.S. Drzaic, *J. Appl. Phys.*, **60**, 6 (1986).
112. K. Toriyama and D.A. Dunmur, *Mol. Cryst. Liq. Cryst.*, **139**, 123 (1986).
113. T. Scheffer and J. Nehring, *J. Appl. Phys.*, **58**, 3022 (1985).
114. E.P. Raynes, *Mol. Cryst. Liq. Cryst. Lett.*, **4**, 1 (1986).
115. E.P. Raynes, *Mol. Cryst. Liq. Cryst. Lett.*, **4**., 69 (1986).
116. S. Saito, M. Kamihara and S. Kobayashi, *Mol. Cryst. Liq. Cryst.*, **139**, 171 (1986).
117. F. Leenhouts and M. Schadt, *J. Appl. Phys.*, **60**, 3275 (1986).
118. M. Schadt and F. Leenhouts, *Appl. Phys. Lett.*, **50**, 236 (1987).
119. C. Waters, E.P. Raynes and V. Brimmell, *Mol. Cryst. Liq. Cryst.*, **123**, 303 (1985).

4 Materials requirements for smectic liquid crystal displays

D. Coates
Standard Telecommunication Laboratories

4.1 Laser addressed smectic A displays

Laser energy was first[1] found to produce a texture change in thin layers of nematic (N) and cholesteric (Ch) liquid crystals in 1972; these initial experiments were soon extended to smectic A (S_A) liquid crystals[2]. The most notable advantages[3] of using an S_A phase (above which there is a narrow-range N phase) are that (1) due to the smaller size and higher density of scattering centres, the light scattering is more intense and of sharper resolution[4] and (2) erasure of written areas is possible.

The basic device usually begins with a clear state which is produced by aligning the liquid crystal either (1) homeotropically[5], (2) homogeneously[6] or (3) as a hybrid of the two[7]. There is conflict of opinion as to which type of alignment should be used; the relevant arguments are given in the cited references. When a focused laser beam is impinged on the thin liquid crystal film (typically 12–14 μm), and provided that the beam can be absorbed, the liquid crystal will become locally heated (Figure 4.1).

The liquid crystal must be heated sufficiently to give the isotropic liquid,

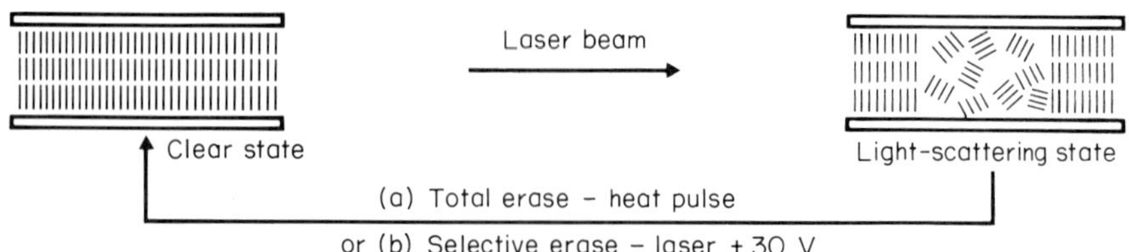

Figure 4.1 The initial clear state is converted into an optically opaque state by the laser energy. Reversal can be achieved either totally over the entire cell or over selected areas.

whereupon subsequent rapid cooling by the heat-sinking action of the glass substrate regenerates the S_A via the N phase; a light-scattering texture is produced. Writing speeds of several centimetres per second are typical. The texture has indefinite storage if the ambient temperature of the cell does not come too close to the S_A–N transition (T_{S_A-N}). The image also has sharp contrast and high resolution which makes it ideal for projection systems[8, 9] where a contrast ratio of 20:1 for the projected image has been reported[10]. The types of projection system used have been discussed by Dewey[11]. It may also be possible to use the effect to produce an optical storage device.

Total erasure is usually effected either by heating all the display to the isotropic liquid by applying a short d.c. pulse to one of the conducting plates, followed either by slow cooling or, more usually, by applying a small voltage while the material in the cell is cooling. This small voltage is sufficient to align the intermediate N phase and produce non-light-scattering alignment (homeotropic) of the subsequent S_A phase. Alternatively, a larger voltage (100–140 V at 1–1.5 kHz) could be applied; this aligns the S_A phase without heating.

Selective erasure is achieved by passing the laser beam over the areas to be erased, and as the material cools, a small electric field (*ca.* 30 V) is applied; this aligns only the N phase which in turn causes the subsequent S_A phase to align homeotropically.

Grey scale can be achieved by applying a field of 5–15 V while the cell is cooling[10]. It has also been suggested that if *ca.* 1 V is applied during cooling, the scattering is more intense[7].

As an alternative mode, the whole cell can be converted into a scattering state either by using light from a flash gun[12] or electrically[13]. Information is then written[14] by applying *ca.* 30 V while a laser beam is passed over the desired areas. This produces clear lines on a scattering background.

Several material properties are essential to produce this display and these are now considered.

4.1.1 Liquid crystal material

The liquid crystal must have an S_A phase and both T_{S_A-N} and T_{N-I} should occur over a narrow temperature range. If the image is to be stored at room temperature, the melting point of the material (usually a mixture) must be below

room temperature so that crystallization does not occur. The S_A phase should also exist over a reasonably wide temperature range. A short N phase is also necessary as this assists in aligning the S_A phase, and allows the display to be 'written' faster by the laser beam[3]. The N phase should be short so that a minimum of laser energy is needed to produce the isotropic phase.

For the same reason it is not desirable to have $T_{S_A\text{–}N}$ too high; however, it must not be too close to ambient temperature or the influence of the aligning layer will dominate and the display's memory will be insecure. Therefore, it is necessary either to thermostat the display at an optimum temperature below $T_{S_A\text{–}N}$, so that external temperature fluctuations do not affect the display, or to provide a means of supplying or absorbing more of the laser energy so that the display can be held well below $T_{S_A\text{–}N}$.

To facilitate erasure (by applying a field), the molecules must have a positive dielectric anisotropy ($\Delta\varepsilon$). To accomplish this, a strong dipole moment along the molecular long axis is required; classically a cyano group is used as one terminal group.

The liquid crystal needs to be stable to high temperatures[15] and high levels of light energy, not just because of the laser light but also because of the intense visible light needed in most projection applications. These energy sources, together with small amounts of water (a few ppm) and possibly oxygen within the cell, mean that the material must be very stable to hydrolysis and

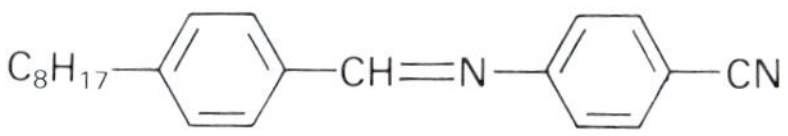

C_8H_{17}–⟨benzene⟩–CH=N–⟨benzene⟩–CN

1

oxidation. Early experiments were carried out using 4-cyano-4′-octylbenzylideneaniline (**1**) and some of its homologues as eutectic mixtures[2, 3]. A typical mixture had the transition temperatures C–S_A, 24 °C; S_A–N, 73.5 °C; N–I, 74 °C — note the narrow N range. However $T_{C\text{–}S_A}$ is quite high, and although the S_A phase supercooled well below room temperature, there was a risk of crystallization. It is also probable that these systems were unstable due to

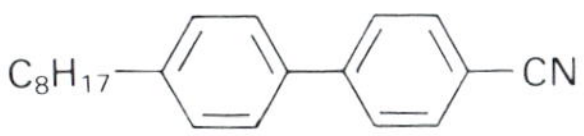

C_8H_{17}–⟨benzene⟩–⟨benzene⟩–CN

2

the hydrolytically sensitive azomethine linkage. They were later replaced by mixtures of 4-alkyl-4′-cyanobiphenyls, e.g. 4-cyano-4′-octylbiphenyl (8CB) (**2**), either alone[10] or as components of eutectic mixtures[16]. Gray and Mosley[17] reported a mixture of 8CB and 4-cyano-4′-decylbiphenyl (10CB) which has the constants C–S_A, 5 °C; S_A–N, 40 °C; N–I, 43 °C. These materials are stable, and are therefore used extensively in mixtures for this and many other applications.

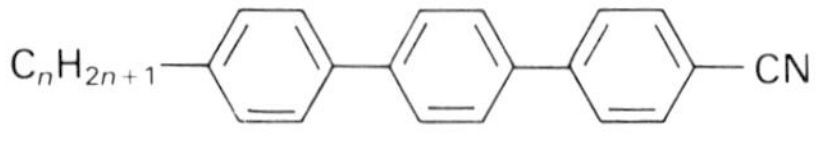

C_nH_{2n+1}—⟨phenyl⟩—⟨phenyl⟩—⟨phenyl⟩—CN

3

$C_nH_{2n+1}O$, $C_nH_{2n+1}O$—⟨phenyl⟩—COO—⟨phenyl⟩—⟨phenyl⟩—CN (or Br)

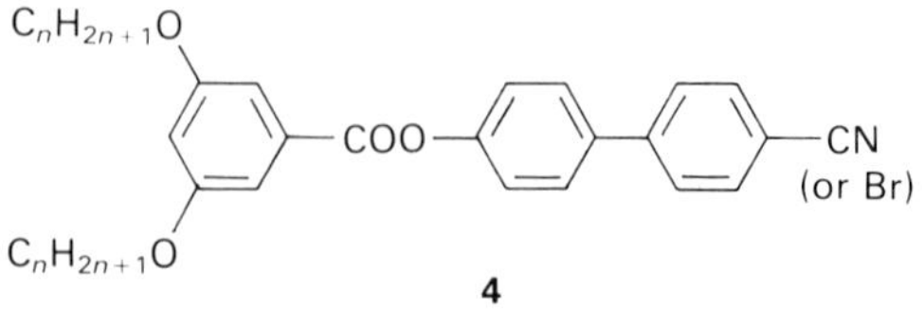

4

When mixed with the analogous 4-alkoxy-4′-cyanobiphenyls and 4-alkyl-4″-cyano-*p*-terphenyls (**3**), S_A phases of very wide temperature range can be made. These mixtures are commercially available[16]. The width of the N phase in mixtures has been reduced by the addition[18] of biphenylyl benzoates (**4**).

Little work has been reported on how the liquid crystal material affects the scattering texture, but it has been shown[19] that a mixture of 8CB and 10CB gives a better scattering texture than the S_A mixture made from 4-heptylphenyl 4-butylbenzoate and 4-cyanophenyl 4-heptylbenzoate.

A slightly different type of display was reported by Tani and Ueno[20], who showed how a coloured display could be made.

4.1.2 Additives to the liquid crystal

Originally the laser energy was absorbed by using an infrared absorbing layer such as indium–tin oxide, which also acted as a transparent (to visible light) conductor to allow application of the electric field. As only a small proportion of the energy is effective in heating the liquid crystal, due mainly to dissipation of the heat by the glass surfaces, a bulky YAG (yttrium aluminium garnet) laser was needed to produce the high power needed. The incorporation of a laser beam absorbing dye (or pigment) in a polymer coated on the cell's inner glass surfaces is possible, but this suffers from the same disadvantages as the indium–tin oxide layer.

A dye dissolved in the liquid crystal could halve[15] the energy needed and reduce spreading of the scattered line. The ideal dye should:

1. Have an absorption maximum coincident with the laser emission wavelength.
2. Be very absorbing, i.e. have a high extinction coefficient so that only a small amount of dye is needed.
3. Be soluble in the liquid crystal.
4. Have a high transmittance at the projection wavelength so that the projection beam is not attenuated.
5. Be very stable to heat and light.
6. Be dichroic, since this helps to reduce the spot width and give better resolution.

He–Ne lasers emit at 0.633 μm, and one series of stable, soluble dyes (**5**) has been reported[21] for this use. The order parameter of these dyes is about 0.6 (order parameter is a measure of the ability of the dye molecule to align with the liquid crystal director). Other blue dyes of higher order parameter (**6** and **7**) have

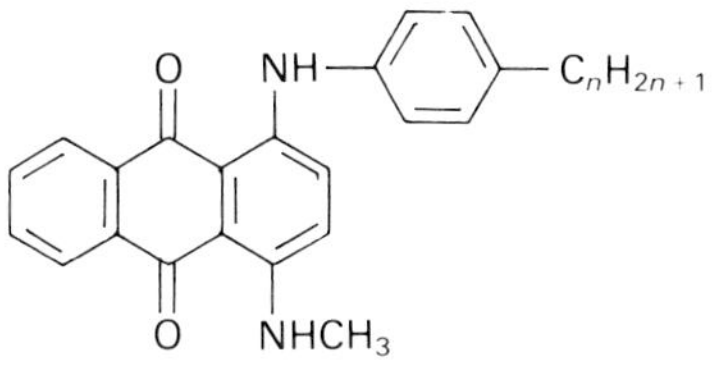

5

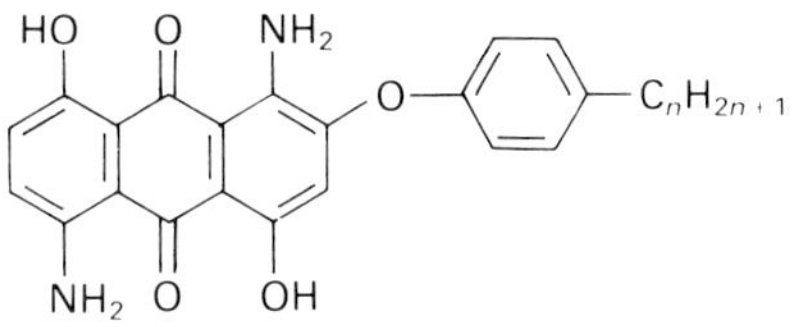

6

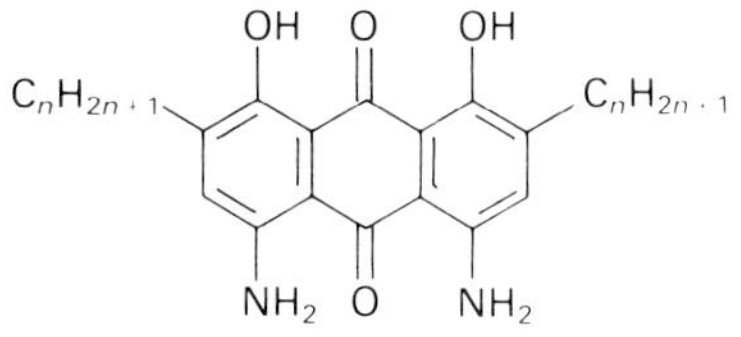

7

been developed[22, 23] for other types of display, but have not been specifically reported for use in this device. Recently, a pyryllium-based cyanine perchlorate dye has been reported[24] which can be used to absorb He–Ne laser energy.

The disadvantage of using a He–Ne laser is its relatively large size and the restriction on the projection light that can be used because the dye is blue.

Semiconductor lasers (GaAs and GaAlAs) seem ideal for this application as they are small, relatively inexpensive and emit in the near-infrared (0.75–1.2 μm), but there are not many suitable dyes that absorb in this region. *N*-Phenyl-*o*-phenylenediamine nickel[25] absorbs the light from a GaAs laser and gives only a faint blue cast to the cell, but it has only a low extinction

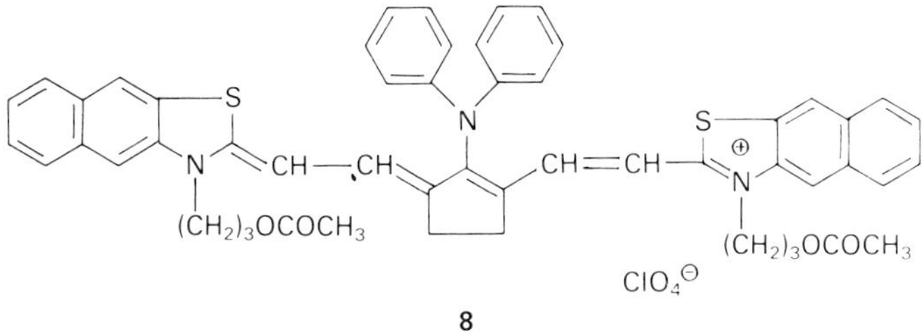

8

coefficient. Some cyanine dyes, such as[26] the cyanine perchlorate (**8**), absorb at about 0.83 μm. Unfortunately, dyes of this type tend to be unstable to light, not very soluble in liquid crystals and (because they are ionic) allow a current to flow in the cell when a voltage is applied; this may cause degradation of the display.

Bis-dithio-α-dicarbonyl complexes[27, 28] absorb in the region 0.7–0.925 μm, depending on the transition metal used and the nature of the aromatic groups. They have high extinction coefficients, but are not particularly light stable or soluble. Giroud *et al.*[29] reported some related complexes (**9**) which exhibit liquid crystal phases. Very recently, an interesting dichroic dye (**10**)

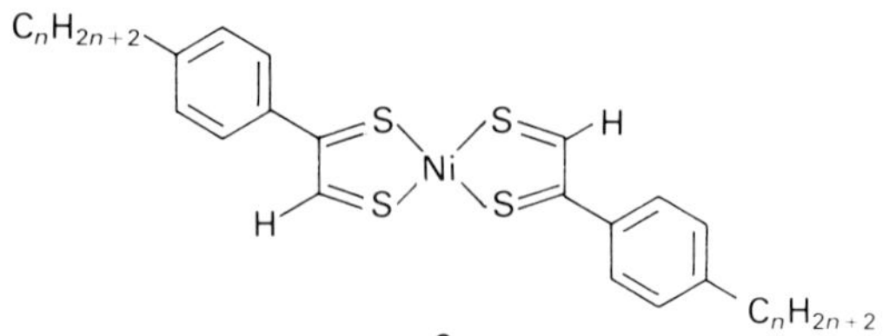

9

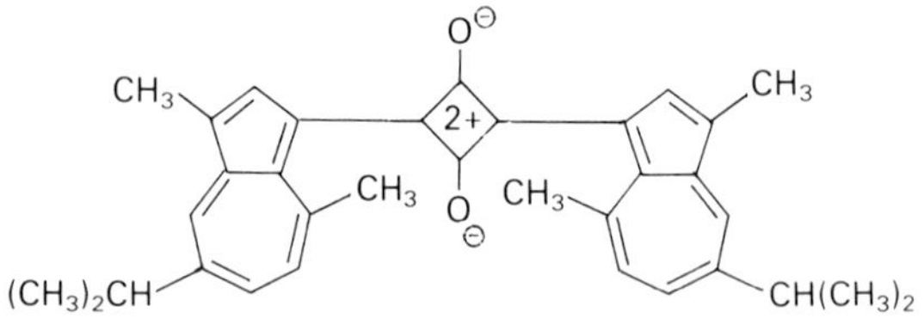

10

was reported[30] with a maximum absorption at 0.77–0.78 μm, corresponding to the emission wavelength of a GaAlAs laser at 0.78 μm.

The incorporation of a dye is also claimed[7] to improve the contrast ratio of the display. The contrast ratio can also be improved by the use of contrast improvement dopants (CIDs), which may also reduce the energy needed to write the scattering texture. Taylor and Kahn[3] claim that non-linear molecules such as adamantane or the *ortho* isomer of the normally *para*-substituted liquid crystal material are good CIDs. However, the effectiveness of such compounds has not been universally claimed[15].

The addition of chiral compounds, such as cholesteryl nonanoate, which convert the N phase into a Ch phase, produces dye-doped cells with faster response and better light scattering[31]. However, the use of cholesteric additives is not without disadvantages[8].

4.2 Thermally/electrically addressed smectic A displays

This effect is very similar in concept to the laser addressed display, but the selective heating of the liquid crystal to the isotropic liquid is achieved using a heat pulse.

The display consists[33, 34] of two pieces of glass about 12 μm apart; the 'back' plate carries rows of evaporated aluminium which act as both a light reflector and as a means of heating the cell along discrete lines. The front glass carries many lines of a transparent conductor (indium–tin oxide). The inner surfaces are coated with a silane-based homeotropic aligning agent. After a heat pulse has been supplied to a line via the aluminium conductor, it is allowed to cool quickly; if no field is applied then a scattering texture will arise. However, if a small field is applied as the liquid crystal cools, the N phase will be homeotropically aligned, thus causing the S_A phase to align similarly; the pixel then appears transparent. Twenty-five rows can be addressed[34] in 2.5s. The image is usually projected. Optimizing the duration and magnitude of the heat pulse with respect to the glass and cell thickness is important[35].

The liquid crystal must exhibit a short N phase and be of positive $\Delta\varepsilon$; the only mixture reported[33, 34] consisted of 8CB and 10CB with compound **11**. The addition of compound **11** reduced the dielectric anisotropy from about 8.4 to 5.6; the transition temperatures were C–S_A, 6 °C; S_A–N, 53 °C; N–I, 56 °C.

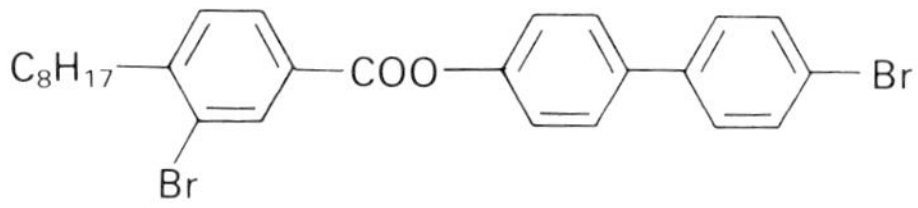

C_8H_{17} — COO — Br
Br

11

This device has been developed[36] into a faster, small area (10 × 12 mm) matrix (256 × 256 lines) display with a line access time of 64 μs, thus allowing video frame rates to be achieved with 32 grey levels.

A development of this basic effect is the TADD[37, 38] (thermally addressed dye display). When the S_A phase is aligned homeotropically, the high-order parameter dye lies parallel to the optic axis and appears almost colourless. In the focal-conic texture, the dye is oriented in a more light-absorbing mode and appears coloured. Thus the display can be viewed directly and produces a black or coloured image on a clear or white background. To increase the speed of the display[39], it is operated at 45 °C; 250 rows can be addressed in 1.25s. No details about the dyes have been reported, but the liquid crystals have a similar requirement to those used in the non-dyed and laser addressed displays, and a mixture of 8CB, 10CB and the chiral additive (*S*)-4-(2-methylbutyl)-4′-cyanobiphenyl has been patented[40].

A second type of S_A device[41] uses liquid crystals of negative $\Delta\varepsilon$. The molecules are initially homeotropically aligned by the use of aligning agents. On applying an electric field, the molecules attempt to lie parallel to the glass, but a degenerate light-scattering orientation results. Erasure is produced by applying a heat pulse, followed by slow cooling from the isotropic phase via a short-range N phase into the S_A phase, which due to surface forces aligns homeotropically. The liquid crystals[42] described for use in this effect are mixtures of 4-alkylphenyl 4-alkoxy-3-cyanobenzoates and laterally substituted phenyl benzoyloxybenzoates. One eutectic mixture has the constants C–S_A, 18 °C; S_A–N, 72 °C; N–I, 80 °C, with a dielectric anisotropy of −4.6.

4.3 Electrically addressed smectic A displays

In 1971, Tani[43] showed that the S_A phase of 4-cyano-4′-octylbenzylideneaniline could be induced into a light-scattering texture, and Hareng *et al.*[44] later showed that the focal-conic or planar texture (molecules essentially parallel to the glass surfaces) of 8CB could be changed into a homeotropic texture.

In 1978, Coates *et al.*[13] demonstrated a totally electrically reversible smectic display effect. This consisted of a clear state and a light-scattering state, both of which had indefinite storage.

At that time S phases were not regarded as serious competitors to N and Ch phases, mainly on the erroneous basis that such viscous phases were bound to have very slow response times. The potential of displays based on S phases was therefore not widely accepted.

The conditions for the formation of the homeotropic (clear) state, by dielectric orientation, were established[13] for 8CB (typically 65 V at 1.5 kHz in an 18 μm thick cell).

The formation of the scattering texture was more complex[13]. For many materials the scattering threshold voltage (V_h) was 50–80 V at low frequencies. To produce a scattering texture, the liquid crystal resistivity had to be below 1×10^9 Ω cm; this was achieved by adding cetyltrimethylammonium bromide, which acted both as an aligning agent and an ionic dopant. The light-scattering effect is created by ions flowing through the S medium under the influence of a low-frequency a.c. field (20–100 Hz). The easy flow path for the ions is along the molecular layers, but occasionally the ions break through the layers. The ratio of this conductivity anisotropy $\sigma_\perp:\sigma_\parallel$ is usually[45] about 3:1. The ionic motion is seen as a turbulent light-scattering texture, and on removal of the field the scattering texture remained.

Materials of higher positive $\Delta\varepsilon$ gave lower threshold voltages; thus 8CB ($\Delta\varepsilon$ = 8.5) had a lower scattering voltage than a mixture of 8CB and 4-n-octyloxyphenyl *trans*-4-n-butylcyclohexane-1-carboxylate ($\Delta\varepsilon$ = 4.5).

As evidence that electrically addressed S displays are viable, this effect has been developed into a complex display having 330 000 picture elements[46]. The storage capabilities of the S_A phase mean that once information has been written, it need not be refreshed. Also, unlike devices using polarizers, there are no viewing angle problems and it is flicker-free. A page of information (420 lines) can be written in about 0.8 s at 15 °C.

The S_A temperature range was −10 to 72 °C (with an operating range of 15–55 °C); above this was an N phase, although it is probably not essential. The mixture is based upon 4-alkyl-4′-cyanobiphenyls to which are added ionic dopants and redox agents which prevent decomposition of the liquid crystal.

An S_A liquid crystal whose sign of $\Delta\varepsilon$ changes at moderate frequency has been reported by Coates[47]. In 4-n-pentyl 2′-chloro-4′(6-n-hexyl-2-naphthoyloxy)benzoate this 'crossover' frequency was 4 kHz. Between crossed polarizers, at frequencies >4 kHz, an activated area appears dark, but at lower frequencies the active area appears birefringent. Therefore, in principle, an electrically reversible, field effect S_A device is possible.

4.4 Ferroelectric liquid crystal displays

The term ferroelectric was for many years applied exclusively to certain types of solids, and the conditions required for their formation are well known[48]. In 1975, Meyer *et al.* reported[49] the symmetry arguments for the existence of a ferroelectric smectic liquid crystal and proved this by synthesizing a compound that exhibited such a phase. The symmetry arguments, initially put forward by Meyer, have since been elaborated[50], and the following conditions must be met:

1. A laminar structure.
2. The molecular long axis must be tilted within the layers, which will therefore possess monoclinic symmetry.
3. The molecules must possess a transverse dipole moment.
4. The molecules must be chiral.

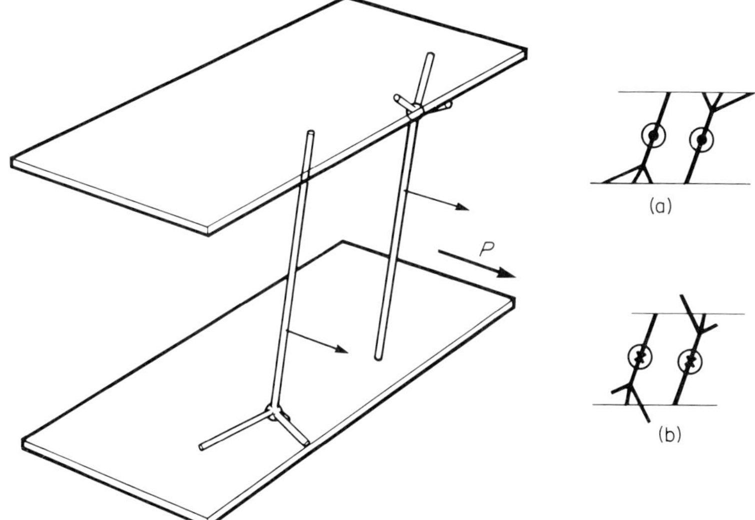

Figure 4.2 Schematic arrangement of chiral molecules in the $S_C{}^*$ phase. Due to the chiral group, depicted by the tripod, the molecules prefer (a) rather than (b) and thus all the molecular dipole moments lie in the same direction.

Various similar models for the ferroelectric liquid crystal phase have been suggested[50–52], and in one[52], the molecules are considered as being tilted within layers and randomly distributed head to tail (Figure 4.2). The chiral centre is depicted as an uneven tripod, with the result that the molecules prefer to tilt in one direction (Figure 4.2a rather than Figure 4.2b) and all the transverse dipoles preferentially lie in one direction, orthogonal to the tilt direction. (The spontaneous polarization (P_s) is a measure of the strength of this dipole moment.) In reality the molecules are spinning rapidly, and this 'preferred' tilt direction then becomes a more energetically favoured position.

The chiral interactions across the layer planes and the reduced symmetry of the phase dictate that the tilt direction will twist through successive layers, thereby giving a helicoidal structure in which the dipole moments average to zero[49];

pitch lengths in the micrometres region are typical. The S_C^* phase is really helielectric, but a truly ferroelectric phase (S_x^*) has been claimed[53].

All tilted S phases[54] should be ferroelectric[50], i.e. the other tilted smectic liquid crystal phases (I* and F*), and the crystal smectic phases (G*, H*, K* and J*). In crystal S phases where the helicoidal forces tend to be overcome by crystal forces, the twist is reduced or eliminated[55] but the dipole moments still average to zero as the phase consists of many randomly oriented monodomains; these have to be aligned before the phase becomes ferroelectric.

Although investigations into the properties of the ferroelectric phase were performed[56–62], it was of little technological interest until 1980 when Clark and Lagerwall[63] demonstrated an electrooptical effect (Figure 4.3) that utilized the ferroelectric property. Potentially the device has the unique combined properties of fast response speed (microseconds as opposed to the usual milliseconds) and bistability. These properties offer the promise of a fast display capable of highly complex functions.

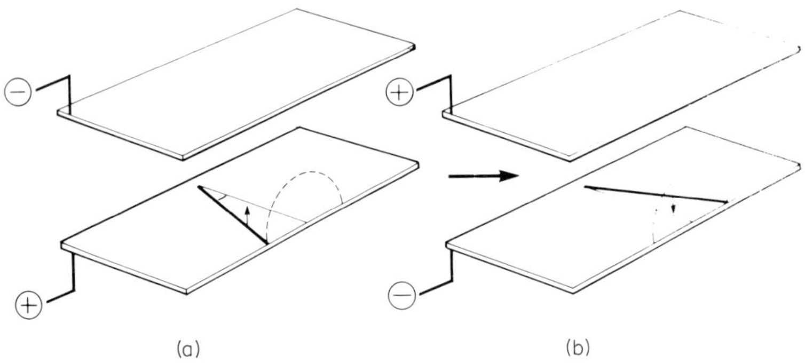

Figure 4.3 When a d.c. voltage is applied across the cell the spontaneous dipole moment (– → +) orientates in the field. To do this, the molecule passes through a half-cone of angle 2θ, where θ is the tilt angle of the molecules within the smectic layers.

The device requires that the molecules are weakly aligned parallel to the glass surfaces[64]; the polarization direction is then perpendicular to the glass plates. By applying a small d.c. bias across the plates, the polarization direction can be changed from up to down or vice versa: this inevitably causes the molecules to undergo a directional change through an angle 2θ. This change is observable between crossed polarizers and is an optimum if the angle 2θ is 45°, i.e. a molecular tilt angle of 22.5°. The two orientation states are energetically of equal stability with an energy barrier between them. It is desirable to encourage the molecules not to relax away from the position into which the applied field has put them — thus ensuring bistability when the field is removed. It is also desirable to unwind the inherent helix of the S_c^* phase. Both these features can be achieved by maximizing the effects of the surface forces; thus very thin cells (1–3 μm thick) are usually used. It is suggested that the speed of the display may be given by $\tau = \eta/P_s E$, where P_s = spontaneous polarization, η = rotational viscosity and E = applied field.

Methods to produce the required alignment have been suggested[64], but there are problems in making uniform thin cells. Some drive schemes have been devised[99, 100] and many new materials developed.

Most S_C-forming compounds do not possess a strong polarity and are therefore of low $\Delta\varepsilon$. For this device, negative $\Delta\varepsilon$ is desirable, as the applied field will then cause the molecules to assume a homogeneous alignment. A method of obtaining bistability in more easily made thick cells has been suggested[65], by using compounds of significant negative $\Delta\varepsilon$; this may also be useful in thin cells incorporating less viscous materials. Once oriented by the d.c. field acting on the ferroelectric dipole, the molecules are held in that orientation by an a.c. field acting on the dielectric dipole. Another option[66] may be to use the $S_I{}^*$ or $S_F{}^*$ phase.

The first reported[67, 99] complex displays based on ferroelectric S_{C^*} liquid crystals had line address times of about 400 μs. More recently line address times at videoframe rates (64 μs) have been reported[101].

4.4.1 *Influence of molecular structure on ferroelectric properties*

Ferroelectric liquid crystal phases are more complex than other liquid crystal phases, and many more parameters, some of which (e.g. elastic constants) have barely been investigated, are required to describe them. However, some factors have been studied and are now reviewed. Changes in molecular structure in particular affect the mutual interactions of the molecules and thereby the properties of the phase.

(a) Tilted phase formation. The major requirements to produce a ferroelectric phase are that the molecules are chiral and tilted within the layers. Several theories[68] have been suggested for the formation of the S_C phase, but none seems to be totally satisfactory. However, three features are common to most materials which form $S_C{}^*$ phases[69]:

1. An aromatic core having two terminal alkyl chains (although this is not always necessary)[70]
2. Strong terminal lateral dipoles
3. At least two aromatic rings

For the phase to be ferroelectric it must also have a chiral centre.

$C_{10}H_{21}O-C_6H_4-CH{=}N-C_6H_4-CH{=}CHCOOCH_2\overset{*}{C}H(CH_3)C_2H_5$

12

The first ferroelectric phase was found[49] in (*S*)-4-n-decyloxybenzylideneamino-2′-methylbutyl cinnamate (DOBAMBC) (**12**), whose transition temperatures are C–$S_C{}^*$, 76 °C; ($S_C{}^*$–$S_I{}^*$, 63 °C); $S_C{}^*$–S_A, 95 °C; S_A–I, 117 °C. Other homologues have since been synthesized[71].

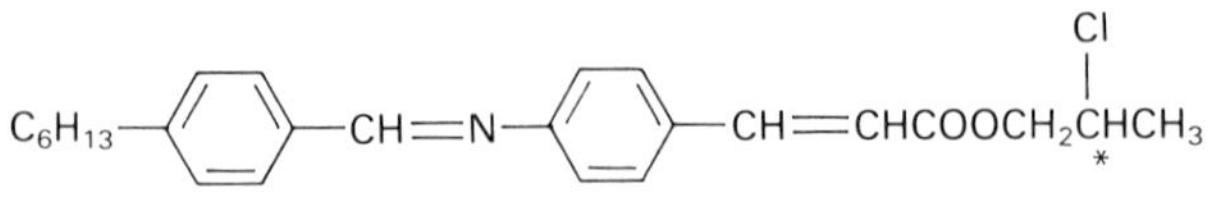

13

Based upon this system (*R*)-4-n-hexyloxybenzylideneamino-2′-chloropropyl cinnamate[72] (HOBACPC) (**13**) was made; its S_C* phase range is shorter (C–S_I*, 65 °C; S_I*–S_C*, 74.5 °C; S_C*–S_A, 81 °C; S_A–I, 136 °C), but the P_s is much higher than that of DOBAMBC[73].

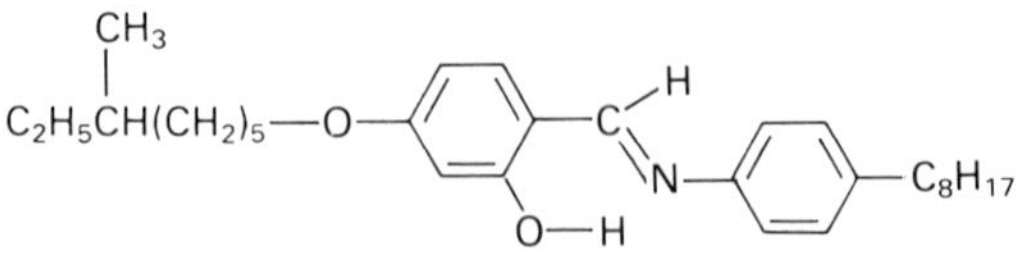

14

In an attempt to produce a room temperature S_C* phase, simpler Schiff's bases were made of which the Mora compounds[74] are notable: the n-octyl homologue (**14**) has the constants C–S_C*, 12 °C; S_C*–I, 97 °C. The P_s of these compounds is extremely low and the tilt angle[75] is only 8°.

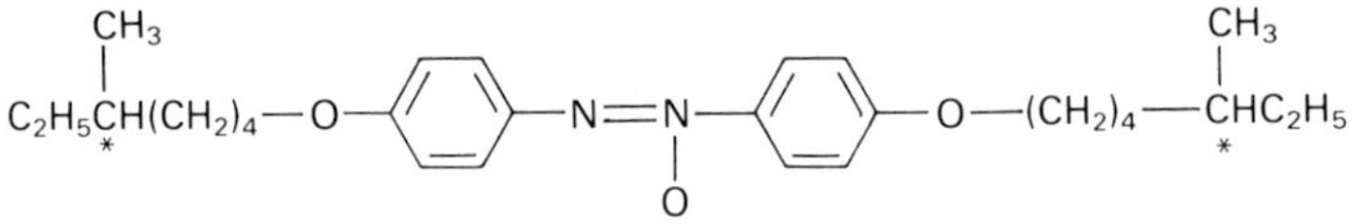

15

The more chemically stable azoxy compounds (**15**) also form S_C phases. The low melting point of **15** (C–S_B, 24 °C; S_B–S_C*, 66 °C; S_C*–Ch, 79.3 °C; Ch–I, 84.2 °C) is spoilt by the introduction of the orthogonal S_B phase[76].

A more systematic study of the occurrence of S_C phases has been made, mainly for ester systems, and some of these results have been summarized[77]. For example, chain branching has a large effect on liquid crystal properties[102], and moving the branching centre away from the

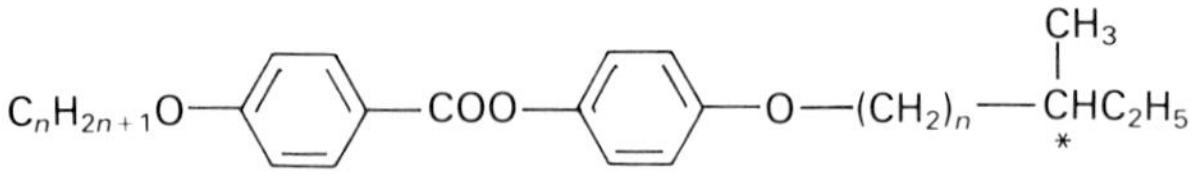

16

central conjugated core often, but not always, increases $T_{S_C-S_A}$, e.g. as in compound **16**, where an increase of as much as 15 °C per carbon atom is found[77]; the effect on the T_{S_C-N} is usually much smaller and occasionally a decrease is found.

Increasing the number of carbon atoms after the branching centre often reduces $T_{S_C-S_A}$ or T_{S_C-N}, e.g. 1-methylheptyl lowers the S_C tendency relative to 1-methylpropyl[78, 79].

The incorporation of small polar lateral substituents can increase the thermal stability of the S_C phase[103].

By increasing the conjugated core using vinyl groups[80] or further phenyl rings[77], most of the liquid crystal phase transition temperatures are increased, but this often results in higher ordered smectic phases, e.g.

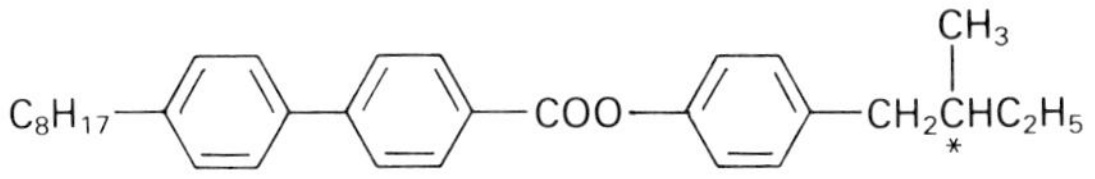

C_8H_{17}–C₆H₄–C₆H₄–COO–C₆H₄–$CH_2\overset{*}{C}H(CH_3)C_2H_5$

17

(*S*)-2-methylbutyl 4′-octylbiphenyl-4-carboxylate[81] (**17**) (C–S_I^*, 66 °C; S_J^*–S_I^*, 65 °C; S_I^*–S_C^*, 69.5 °C; S_C^*–S_A, 84 °C; S_A–Ch, 135 °C; Ch–I, 141 °C. A similar result is achieved if two ester linkages are used, although this tends to increase S_A more than S_C properties[77].

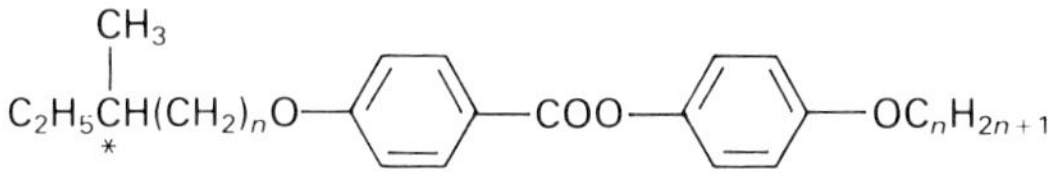

$C_2H_5\overset{*}{C}H(CH_3)(CH_2)_nO$–C₆H₄–COO–C₆H₄–$OC_nH_{2n+1}$

18

$C_2H_5\overset{*}{C}H(CH_3)(CH_2)_nO$–C₆H₄–OOC–C₆H₄–$OC_nH_{2n+1}$

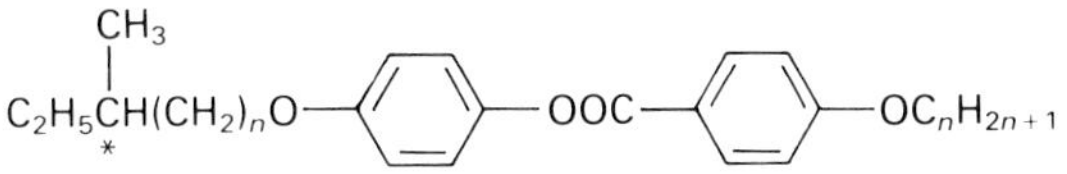

19

Reversing the ester linkage can have a dramatic effect upon the phase sequence[69, 77], e.g. the compounds **18** show the sequence C–S_C^*–S_A–I while compounds **19** show the sequence C–S_C^*–Ch–I. On the other hand, in another

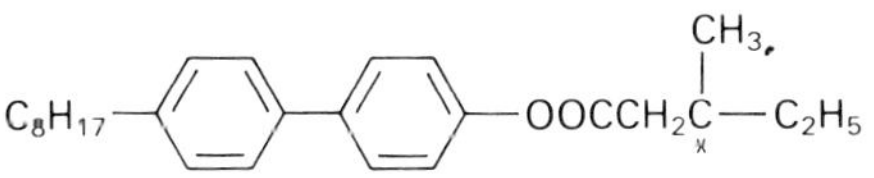

C_8H_{17}–C₆H₄–C₆H₄–$OOCCH_2\overset{*}{C}H(CH_3)$–$C_2H_5$

20

$C_8H_{17}O$–C₆H₄–C₆H₄–$COOCH_2\overset{*}{C}H(CH_3)C_2H_5$

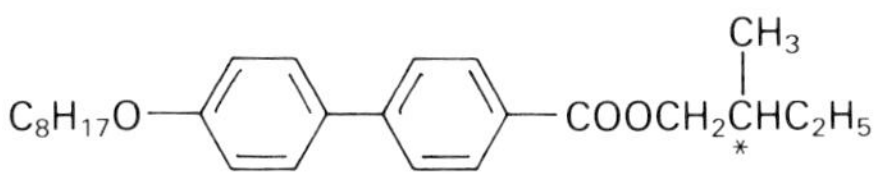

21

system, the type of tilted phase changes[69], e.g. **20**, exhibits T_{S_G-I}, while **21** exhibits $T_{S_C-S_A}$.

Derivatives of pyrimidine[67], biphenylylcyclohexanecarbonitriles[104] and benzalazines[82] also exhibit S_C* phases.

The relation between molecular structure and the occurrence of S_I and S_F phases is less well studied.

It is advantageous to incorporate an S_A phase in the compound or mixture as this is more easily aligned and helps align the S_C* phase[64].

(b) Tilt angle. Several electrooptical methods of measuring the tilt angle of ferroelectric phases have been described[56], but results (ranging from 8 to 45°) have not yet been related to molecular structure as few compounds have been studied. The change in tilt angle (ideally 22.5° for the Clark–Lagerwall device) is often large near $T_{S_C-S_A}$ or T_{S_C-N} (Figure 4.4)[83].

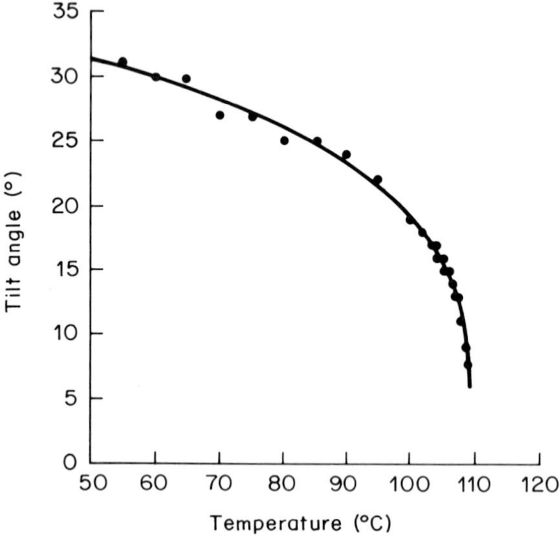

Figure 4.4 Plot of tilt angle versus temperature in the S_C* phase of a proprietary compound STL85.

(c) Helical pitch. The helical pitch sense (*laevo* or *dextro*) in cholesterics has been related[84] to the absolute configuration of the chiral group and the position of the chiral atom relative to the conjugated core of the molecule. This relationship also seems to apply to tilted S phases. However, the original rule appears to be reversed when strongly polar groups are attached to the chiral centre[69]. When the chiral centre is moved stepwise away from the core of the molecule, the twist sense of the helix alternates from *laevo* to *dextro* and becomes longer.

The required long pitch lengths can be obtained by mixing together compounds of opposite helical sense. This has been demonstrated[60] by mixing (*S*)-DOBAMBC and (*R*)-HOBACPC to produce a mixture of very long pitch, but high P_s (15 nC/cm^2). It has also been shown[105–107] that the S_C* and S_A phases are aligned more easily if they are formed via a cholesteric phase of very long or infinite pitch length.

Unlike most cholesterics, the helical pitch of S_C* phases shows a general decrease with decreasing temperature. The S_C* helix can be unwound by surface forces when the cell spacing is smaller than a quarter of the helical pitch, but long pitch S_C* phases may not allow thick cells to be used in the Clark and Lagerwall device, as the thin cells probably also confer a degree of bistability on the aligned states which may not occur with thick cells.

(d) Critical field. This is the energy required to cause the dipole moments to align in the field and is given by the expression

$$E_c = \frac{\pi^4 K\theta}{4P_s Z^2}$$

where Z = pitch length, θ = tilt angle and K = the torsional elastic constant. The magnitude of E_c is usually about 10 V/cm; however, most compounds studied have contained the same chiral group ((*S*)-2-methylbutyl) and one of the few other compounds studied is HOBACPC, which has an E_c an order of magnitude lower than that of DOBAMBC[56].

(e) Rotational viscosity and birefringence. The contrast ratio of ferroelectric displays using polarizers is determined by the molecular tilt angle (θ), the cell thickness (d) and the birefringence (n_a) of the liquid crystal when viewed with light of wavelength λ. When perfect extinction is achieved in one domain (up), the intensity of light passing through the opposite domain (down) is given by $\sin^2(4\theta)\ \sin^2(\pi n_a d/\lambda)$. For a birefringence of 0.25 (at 500 nm) the optimum cell thickness is 1 µm; lower birefringence values would clearly allow thicker cells to be used[85]. If non-optimized thickness/birefringence parameters are used, the device will show a birefringence colour instead of white in its bright state.

Rotational viscosity appears to be a major factor in determining the response times of ferroelectric displays (see Section 4.4), but very few measurements have been reported.

(f) Spontaneous polarization. Spontaneous polarization (P_s) has been measured by various methods[87, 88]; it varies with tilt angle and temperature. A typical plot of P_s versus temperature is shown[83] in Figure 4.5.

If the rotation sign (i.e. *R* or *S*) of the chiral centre is changed, the polarization direction changes by 180°, and so two polarization directions exist[52]. The direction of the permanent dipole can be opposite for different compounds with the same chiral centre (compound **22** has the opposite sign to compound **23**). Consequently, the dipoles in the core of the molecule may influence the sign of P_s[89].

The P_s sign of the (*S*)-2-methylbutyl and (*S*)-3-methylpentyl esters of **24** have, however, been shown to be opposite, and so the P_s sign seems to alternate[69] in the same way as helical pitch sense for homologues.

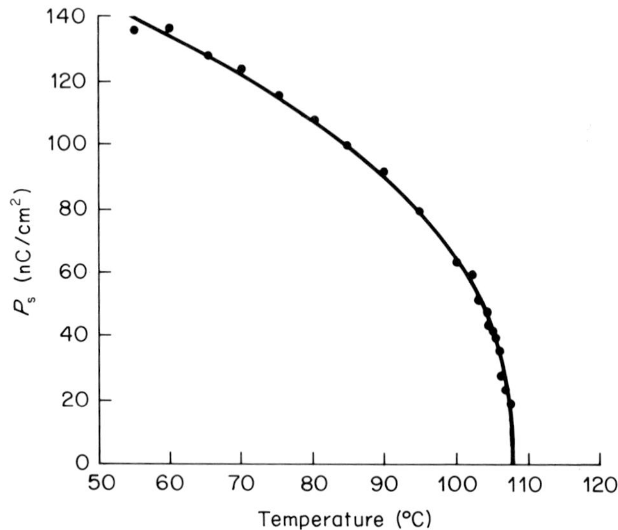

Figure 4.5 Plot of spontaneous polarization (P_s) versus temperature in the S_C* phase of a proprietary compound STL85.

$C_8H_{17}O$–C$_6$H$_4$–CH=N–C$_6$H$_4$–CH=C(Cl)–COOCH$_2$C*H(CH$_3$)C$_2$H$_5$

22

$C_8H_{17}O$–C$_6$H$_4$–CH=N–C$_6$H$_4$–CH=C(CN)=COOCH$_2$C*H(CH$_3$)C$_2$H$_5$

23

$C_{11}H_{23}O$–C$_6$H$_4$–CH=CHCOO–C$_6$H$_4$–COO(CH$_2$)$_n$–C*H(CH$_3$)–C$_2$H$_5$

24

If two ferroelectric liquid crystal compounds are mixed together, the dipoles will either cancel or reinforce, depending on the sign of their polarizations. A convention to describe the two possible P_s signs (+ or −) has recently been suggested[50].

Unambiguously defining the magnitude of P_s is an unsolved problem. The maximum P_s can be very different from the P_s quoted at a specific reduced temperature.

According to the equation in Section 4.4, $\tau = \eta / P_s E$, liquid crystals with large P_s values are needed for the operation of high-speed electrooptic devices at low voltages. The P_s found for liquid crystals is about 10^2–10^3 times smaller than that typically found for solids, and it is also 10^2 times smaller than would be expected from the permanent dipole moments in the molecule[50, 91]. Two effects could be responsible for this: (1) weak coupling of the molecular motion to the monoclinic environment and (2) weak coupling of the chiral group to the polar part of the molecule. Therefore, to obtain a high P_s, the relative orientation of the molecules will be important and the chiral group must be strongly coupled to the molecular core and the polar parts of the molecule. Therefore, the nature of the chiral group will be important, and a strong, rigidly coupled lateral dipole moment should also help.

The following examples show how sterically hindering the rotation of the chiral

$C_8H_{17}O$–(C₆H₄)–(C₆H₄)–COO–(C₆H₄)–$CH_2CH_2\overset{*}{C}H(CH_3)$–$C_2H_5$

25

$C_{10}H_{21}O$–(C₆H₄)–(C₆H₄)–CH=N–(C₆H₄)–CH=CHCOO$\overset{*}{C}H(CH_3)C_3H_7$

26

centre can have a large effect on the P_s. The P_s of compound **25** is[83] 0.14 nC/cm^2, of compound **17** is [92] 3.9 nC/cm^2 and of compound **26** is[73] 40 nC/cm^2. However, things are not this simple, for compound **27** has[73] a P_s of 18 nC/cm^2, which may be due to a difference in the intermolecular arrangements in the phases of compounds **27** and **26**.

$C_{10}H_{21}O$–(C₆H₄)–CH=N–(C₆H₄)–CH=CHCOO$\overset{*}{C}H(CH_3)C_2H_5$

27

Other compounds with 1-methylheptyl terminal groups have been made with P_s values[93] up to 70 nC/cm^2. By enlarging the chiral group to increase steric hindrance, P_s values of up to 70 nC/cm^2 have also been found[94], and the P_s value can also be increased by placing polar groups on the chiral centre; thus HOBACPC has a maximum P_s value of about 20 nC/cm^2 compared[73] to 4 nC/cm^2 for DOBAMBC. If the chiral centre is both sterically hindered and very close to a polar group, very high P_s values of between 80 and 200 nC/cm^2 are found[96]. More recently the same feature has been exploited in the 4′-alkoxy-4-biphenylyl esters of (2S,3S)-3-methyl-2-chloropentanoic acid[109]. These compounds have P_s values of about 250 nC/cm^2.

Recent results[83] indicate that the speed of the Clark and Lagerwall display may not be described by the equation proposed in Section 4.4 and that the viscosity may be very important. Also, the peak current required to switch a ferroelectric liquid crystal with a P_s of *ca.* 100 nC/cm^2 in a 2 μm thick cell is extremely high — beyond the limits of drive circuitry fabricated by conventional technology[95]. However, the need for materials of high P_s is still important, as they will probably be used as dopants (see Section 4.4.2).

Most reports of P_s have been concerned with the $S_C{}^*$ phase. Initial investigation[92] of the more ordered phases showed that an increase in P_s can occur at $T_{S_C{}^*-S_I{}^*}$; the more ordered $S_G{}^*$ phase has, however, a very high E_c and the P_s is not easily measured. However, it has subsequently[83] been found that many compounds can show a slight decrease in P_s at the $T_{S_C{}^*-S_I{}^*}$.

To describe a compound adequately, the absolute configuration (*S* or *R*), helical sense (*laevo* or *dextro*), pitch length, magnitude and sign of P_s, tilt angle and transition temperatures need to be defined; very few compounds to date have been so described.

4.4.2 Ferroelectric mixtures

In 1980 it was shown[97] that an S_C phase doped with chiral molecules, such as L-menthol, gave a ferroelectric phase. This result has since been extended to other systems.

By adding a non-mesomorphic, but structurally liquid crystal-like compound containing a lateral polar group to an S_C or $S_C{}^*$ material (such as DOBAMBC) a large increase in P_s occurs[98]. This is important since the molecular structural requirements to produce a high P_s compound are generally not compatible with requirements to produce a wide temperature range $S_C{}^*$ phase. The work on mixtures shows that a high P_s, wide temperature range, ferroelectric phase can probably be made by mixing a material of high P_s, but not necessarily exhibiting a tilted phase, with a compound, or mixture, having a wide temperature range tilted phase; thus each type of component can be optimized and then mixed together. Several mixtures using this principle are now commercially available[108] and many others have been described in recent patents.

It is common practice to mix several N or S_A compounds together to achieve an average T_{N-I} or T_{S_A-N} but a lowered melting point, thereby widening the temperature range of the phase; the outcome of mixing S_C compounds together is not as simple. In some cases the S_C phase is reduced by a predominant S_A phase or by the introduction of an otherwise absent S_B phase; some examples of this unusual behaviour have been documented[69], but a great deal more work needs to be done.

4.5 References

1. H. Melchior, F.J. Kahn, D. Maydan and D.B. Fraser, *Appl. Phys. Lett.*, **21**, 392 (1972).
2. F.J. Kahn, *Appl. Phys. Lett.*, **22**, 111 (1973).
3. G.N. Taylor and F.J. Kahn, *J. Appl. Phys.*, **45**, 4330 (1974).
4. D.F. Aliev and A.Kh. Zeinally, *Zh. Tekh. Fiz.*, **52**, 1669 (1982).

5. T. Urabe, K. Arai and A. Ohkoshi, *J. Appl. Phys.*, **54**, 1552 (1983).
6. A.G. Dewey, J.T. Jacobs and B.G. Huth, *SID Symp. Dig.*, **1978**, 19.
7. W.H. Chu and D.Y. Yoon, *Mol. Cryst. Liq. Cryst.*, **54**, 245 (1979).
8. J. Harrold and C. Steele, *Eurodisplay Proc. (Paris)*, **1984**, 29.
9. A.G. Dewey, S.F. Anderson, G. Cheroff, J.S. Feng, C. Handen, H.W. Johnson, J. Leff, R.T. Lynch, C. Marinelli and R.W. Schmiedeskamp, *SID Symp. Dig.*, **1983**, 36.
10. M. Hareng and S. LeBerre, *Electron. Lett.*, **11**, 73 (1975).
11. A.G. Dewey, *Inst. Elect. Electron Engrs., Trans. Electron. Devices*, **24**, 918 (1977).
12. Br. Pat. 2,093,206A (1982).
13. D. Coates, W.A. Crossland, J.H. Morrissy and B. Needham, *J. Phys. D, Appl. Phys.*, **11**, 2025 (1978).
14. C.J. Walker and W.A. Crossland, *Displays*, **1985**, 207.
15. D. Armitage, *J. Appl. Phys.*, **52**, 4843 (1981).
16. BDH Ltd, Poole, Dorset.
17. G.W. Gray and A. Mosley, *J. Chem. Soc., Chem. Commun.*, **1976**, 147.
18. US Pat. 4,474,679 (1984).
19. V.I. Andreev, A.N. Nesrullaev, A.S. Sonin and B.M. Stepanov, *Chem. Abs.*, **94**, 200749 (1972).
20. C. Tani and T. Ueno, *Appl. Phys. Lett.*, **33**, 275 (1978).
21. Br. Pat. 2,091,1753A (1982).
22. Br. Pat. 2,069,518 (1981).
23. Br. Pat. 2,081,736 (1982).
24. Jpn. Pat. 1076243 (1985); *Chem. Abs.*, **103**, 113435 (1985); see also *Chem. Abs.*, **103**, 132522–3, 132525, 132527–1327530, and 132533–6 (1985).
25. R. Balanson, W.H. Chu, R.J. Cox and J.T. Jacobs, *IBM Tech. Discl. Bull.*, **21**, 2007 (1978).
26. US Pat. 3,774,122 (1973).
27. G.N. Schrauser and V.P. Mayweg, *J. Am. Chem. Soc.*, **87**, 1483 (1965).
28. G.N. Schrauser and V.P. Mayweg, *J. Am. Chem. Soc.*, **87**, 3583 (1965).
29. A.M. Giroud, A. Nazzal and U.T. Mueller-Westerhoff, *Mol. Cryst. Liq. Cryst.*, **56**, 225 (1980).
30. Br. Pat. 2,140,023A (1984).
31. A. Sasaki, N. Hayashi and T. Ishibashi, *Japan Displays*, **1983**, 497.
32. M. Hareng and S. LeBerre, *Proc. IEDM*, **1978**, 258.
33. S. LeBerre, M. Hareng, R. Hehlen and J.N. Perbet, *Displays*, **1981**, 349.
34. S. LeBerre, M. Hareng, R. Hehlen and J.N. Perbet, *Proc. Electron. Displays*, **1982**, 39.
35. M. Hareng, S. LeBerre, B. Mourey, P.C. Moutou, J.N. Perbet and L. Thirant, *Proc. Displays Res. Conf.*, **1982**, 126.
36. M. Hareng, R. Hehlen, S. LeBerre, J.P. LePesant and L. Thirant, *Proc. Electron. Displays Conf.*, **1**, 28 (1984).
37. *Information Display*, **1**, 3 (1982).
38. S. Lu, D.H. Davies, C.H. Chung, D. Evanicky, R. Albert and R. Traber, *Int. Displays Res. Conf.*, **1982**, 132.
39. W. Harter, S. Lu, S. Ho, B. Basler, W. Newman and C. Otaguro, *SID Symp. Dig.*, **1984**, 196.
40. US Pat. 4,461,715 (1984).
41. J.C. Dubois, *Ann. Phys.*, **3**, 131 (1978).
42. J.C. Dubois, A. Zann and A. Beguin, *Mol. Cryst. Liq. Cryst.*, **42**, 139 (1977).
43. C. Tani, *Appl. Phys. Lett.*, **19**, 241 (1971).
44. M. Hareng, S. LeBerre and L. Thirant, *Appl. Phys. Lett.*, **25**, 683 (1974) and **27**, 575 (1975).
45. L.M. Blinov, *Electro-Optical and Magneto-Optical Properties of Liquid Crystals*, John Wiley and Sons, Chichester (1983).
46. W.A. Crossland and S. Cantor, *SID Symp. Dig.*, **1985**, 124.
47. D. Coates, *Mol. Cryst. Liq. Cryst.*, **49**, 83 (1978).
48. F. Jona and G. Shirane, *Ferroelectric Crystals*, Pergamon Press, New York (1962).
49. R.B. Meyer, L. Liebert, L. Strzelecki and P. Keller, *J. Phys. (Paris) Lett.*, **36**, 69 (1975).
50. S.T. Lagerwall and I. Dahl, *Mol. Cryst. Liq. Cryst.*, **114**, 151 (1984).
51. M.A. Ospirov and S.A. Pikin, *Mol. Cryst. Liq. Cryst.*, **103**, 57 (1983).
52. L.A. Beresnev and L.M. Blinov, *Ferroelectrics*, **33**, 129 (1981).
53. H.R. Brand and P.E. Cladis, *Mol. Cryst. Liq. Cryst.*, **114**, 207 (1984).
54. G.W. Gray and J.W. Goodby, *Smectic Liquid Crystals*, Leonard Hill, Glasgow (1984), p. 153.
55. J. Doucet, P. Keller, A.M. Levelut and P. Porquet, *J. Phys. (Paris)*, **39**, 548 (1978).

56. Ph. Martinot-Lagarde, R. Duke and G. Durand, *Mol. Cryst. Liq. Cryst.*, **75**, 249 (1981).
57. T. Uemoto, K. Yoshino and Y. Inuishi, *Mol. Cryst. Liq. Cryst.*, **67**, 137 (1981).
58. L. Benguigui and F. Hardouin, *J. Phys. (Paris) Lett.*, **42**, L381 (1981).
59. L.J. Yu, H. Lee, C.S. Bak and M.M. Labes, *Phys. Rev. Lett.*, **36**, 388 (1976).
60. S. Garoff and R.B. Meyer, *Phys. Rev. A*, **18**, 2739 (1978) and **19**, 338 (1979).
61. S.A. Pikin and V.L. Indenbom, *Ferroelectrics*, **20**, 151 (1978).
62. W. Kucyinski, *Ber. Bunsenges. Phys. Chem.*, **85**, 234 (1981).
63. N.A. Clark and S.T. Lagerwall in *Liquid Crystals of One- and Two-Dimensional Order*, ed. by W. Helfrich and G. Heppke, Springer Verlag, Berlin, (1980), p. 222.
64. J.S. Patel, T.M. Leslie and J.W. Goodby, *Ferroelectrics*, **61**, 137 (1984).
65. J.P. LePesant, B. Mourey, M. Hareng, G. Decobert and J.C. Dubois, *J. Phys. (Paris)*, **C-5**, 217 (1984).
66. W.A. Crossland, M.F. Bone and A.B. Davey, Br. Pat. 8,607,953.
67. T. Harada, M. Taguchi, K. Iwasa and M. Kai, *SID Symp. Dig.*, **1985**, 131.
68. G.W. Gray and J.W. Goodby, *Smectic Liquid Crystals*, Leonard Hill, Glasgow (1984), p. 45.
69. J.W. Goodby and T.M. Leslie, *Mol. Cryst. Liq. Cryst.*, **110**, 175 (1984).
70. Y. Shimizu and S. Kusabayashi, *Chem. Lett.*, **1985**, 1005.
71. P. Keller, L. Liebert and L. Strzelecki, *J. Phys. (Paris)*, **37**, 3 (1976).
72. P. Keller, S. Juge, L. Liebert and L. Strzelecki, *C.R. Hebd. Seances Acad. Sci.*, **282**, 639 (1976).
73. K. Yoshino, M. Ozaki, T. Sakurai, K. Sakamoto and M. Honma, *Jpn. J. Appl. Phys.*, **23**, L175 (1984).
74. A. Hallsby, M. Nilsson and B. Otterholm, *Mol. Cryst. Liq. Cryst.*, **82**, 69 (1982).
75. K. Sharp, K. Flatischler, K. Kondo, Y. Sato, K. Miyasato, H. Takezoe, A. Fukada and E. Kuze, *Jpn. J. Appl. Phys.*, **22**(4), 566 (1983).
76. P. Keller, *Ann. Phys.*, **3**, 139 (1978).
77. J.W. Goodby and T.M. Leslie in *Liquid Crystals and Ordered Fluids*, Vol. 4, ed. by J. F. Johnson and A.D. Griffin, Plenum Press, New York (1984), p. 1.
78. P. Keller, *Mol. Cryst. Liq. Cryst.*, **102**, 295 (1984).
79. Eur. Pat. 115,693 (1984).
80. T.M. Leslie, *Ferroelectrics*, **8**, 9 (1984).
81. S. Budai, R. Pindak, S.C. Davey and J.W. Goodby, *J. Phys. (Paris) Lett.*, **45**, 1053 (1984).
82. J.L. Alabart, M. Marcos, E. Melendez and J.L. Serrano, *Ferroelectrics*, **58**, 37 (1984).
83. M.F. Bone and D. Coates, unpublished data.
84. G.W. Gray and D.G. McDonnell, *Mol. Cryst. Liq. Cryst.*, **34**, 211 (1977).
85. N.A. Clark, M.A. Handschy and S.T. Lagerwall, *Mol. Cryst. Liq. Cryst.*, **94**, 213 (1983).
86. K. Sarp, I. Dahl, S.T. Lagerwall and B. Stebler, *Mol. Cryst. Liq. Cryst.*, **114**, 283 (1984).
87. J. Hoffman, W. Kuczynski and J. Malecki, *Mol. Cryst. Liq. Cryst.*, **44**, 287 (1978).
88. Ph. Martinot-Lagarde, *J. Phys. (Paris) Lett.*, **38**, L-17 (1977).
89. Ph. Martinot-Lagarde, *J. Phys. (Paris)*, **37**, 129 (1976).
90. J. Billard, *Ferroelectrics*, **58**, 81 (1984).
91. M.V. Loseva, B.I. Ostrovski, A. Rabinovich, A.S. Sonin, B.A. Strukov and N.I. Chernova, *JETP Lett.*, **28**, 375 (1978).
92. M.F. Bone, D. Coates and A.B. Davey, *Mol. Cryst. Liq. Cryst.*, **102**, 331 (1984).
93. Br. Pat. application and Eur. Pat. 110,299 (1984).
94. Br. Pat. PCT/GB/0046 and M.F. Bone, D. Coates, G.W. Gray, D. Lacey, K.J. Toyne and D.J. Young, *Mol. Crys. Liq. Cryst. Lett.*, **3**, 189 (1986).
95. W.A. Crossland, personal communication.
96. Br. Pat. PCT/GB/00512 (1986) and M.J. Bradshaw, M.F. Bone, L.K.M. Chan, D. Coates, J. Constant, P.A. Gemmell, G.W. Gray, D. Lacey, E.P. Raynes, K.J. Toyne and D.J. Young, paper presented at the 11th International Liquid Crystal Conference, Berkeley, California 1986.
97. W. Kuczynski and H. Stegemeyer, *Chem. Phys. Lett.*, **70**, 123 (1980).
98. L.A. Beresnev, V.A. Baikalov and L.M. Blinov, *Sov. Phys. Tech. Phys.*, **27**, 1296 (1982).
99. M. Geary, *SID Digest*, **1985**, 128.
100. Br. Pat. application 86/07952.
101. M.F. Bone, D. Coates, W.A. Crossland, P. Gunn and P.W. Ross, *Displays*, in press.
102. D. Coates, *Liquid Crystals*, **2**, 63 (1987).
103. D. Coates, *Liquid Crystals*, to be published, 1987.

104. R. Eidenschink, R. Hopf, B.S. Scheuble and A.F. Wächtler, Freiburger Arbeitstagung Flüssigkristalle, Freiburg, 1986.
105. M.J. Bradshaw, V. Brimmel and E.P. Raynes, presented at the SID Conference, Japan Display '86, Post Deadline Paper 5, and Br. Pat. application 86/08114.
106. Eur. Pat. 178,647 (1986).
107. US Pat. 4,615,586 (1986).
108. For example, mixtures such as SCE3 from BDH Ltd., Poole, Dorset, U.K. and ZLI 3654 from E. Merck, Darmstadt, FRG.
109. K. Yoshino, S. Kishio, M. Ozaki, T. Sakurai, N. Mikami, R. Higuchi and M. Honma, *Jpn. J. Appl. Phys.*, **25**, 416 (1986).

5 Thermochromic cholesteric liquid crystals

D.G. McDonnell
Royal Signals and Radar Establishment

5.1 Introduction

The cholesteric (Ch) mesophase is closely related to the nematic (N) phase but differs significantly in that the director (the unit vector describing the average direction of the molecular long axes) is not constant in space, but undergoes a helical distortion[1]. The structure may be envisaged as composed of layers or sheets of nematic liquid crystal (NLC), as shown in Figure 5.1. The director of an individual layer is rotated through a small angle with respect to the director in adjacent layers. As a succession of layers is passed through, the director turns through 360° and this thickness represents the pitch length for the helix. It is important to emphasize that the concept of layers or sheets used in this simple pictorial representation has of course no real meaning, but provides a good approximation.

The main chemical feature which distinguishes a Ch material from an N material is that its molecular structure is chiral and therefore not superimposable on its mirror image. The importance of structural optical activity and the special

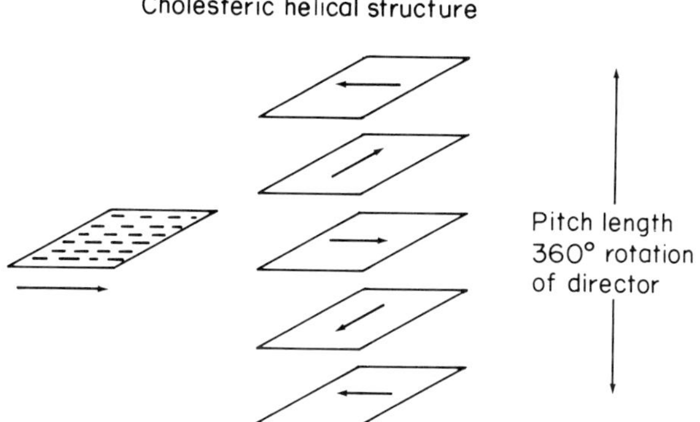

Figure 5.1 Idealized helical model of a cholesteric liquid crystal.

relationship between the N and Ch phases has been demonstrated by two noteworthy observations:

1. N phases are made Ch by dissolving in them optically active enantiomers which themselves may[2] or may not[3] be LC materials.
2. A racemate exhibiting an N phase can be resolved[4] into its enantiomers, each of which will exhibit a Ch phase. A further point to note is that the helical twist senses of the phases formed by the two enantiomers are opposite in sign, indicating a relationship of helical sense to molecular structure[5], and this will be examined in a later section.

The helical structure gives rise to two unique identifying textures when microscope slide preparations of the Ch phase are examined using a polarizing microscope. The first of these, shown in Figure 5.2, is the focal-conic texture which forms on melting or on cooling from the isotropic phase. This is analogous to the S focal-conic texture, but forms as the helices pack around an ellipse and a hyperbola[6]. Since this structure provides a regularly changing distribution of the optic axis, the film, when viewed in ambient light, scatters light and appears opaque. If the cover slip is mechanically disturbed a second texture, the Grandjean[6] plane texture, is formed as the helices flow align with their axes orthogonal to the substrates (Figure 5.3). This texture is particularly important, since it can reflect light selectively, and when viewed in ambient light can exhibit a highly iridescent colour.

Another interesting microscopic feature of the Ch phase often occurs over a narrow temperature interval when the mosaic platelet texture of the 'blue phase' forms from the isotropic phase. Consideration of this curious structure is beyond the scope of the present work, and the reader is referred to the review by Crooker[7].

5.2 Selective reflection and thermochromism

Classically, isotropic liquids or solutions containing chiral molecules, e.g. sugar

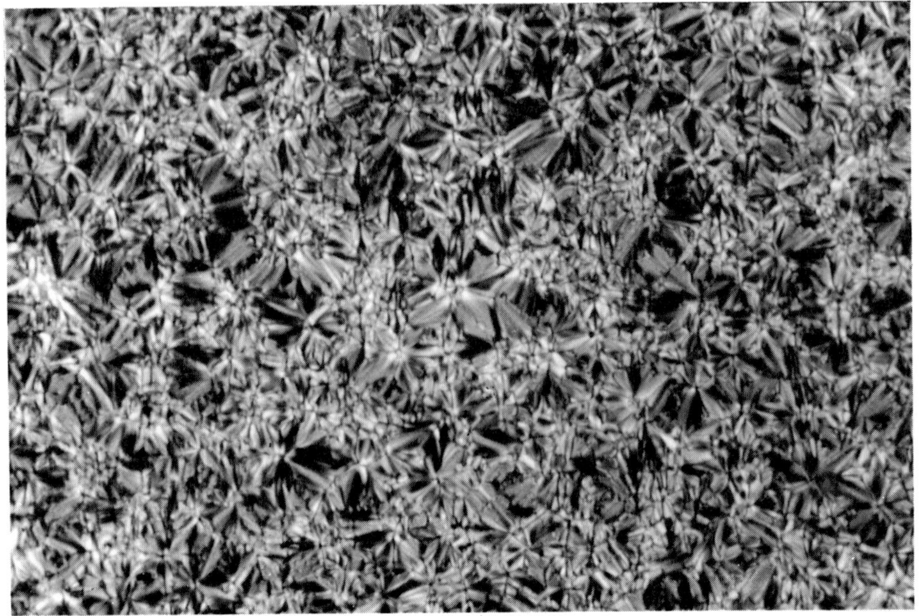

Figure 5.2 Focal-conic texture of a cholesteric liquid crystal.

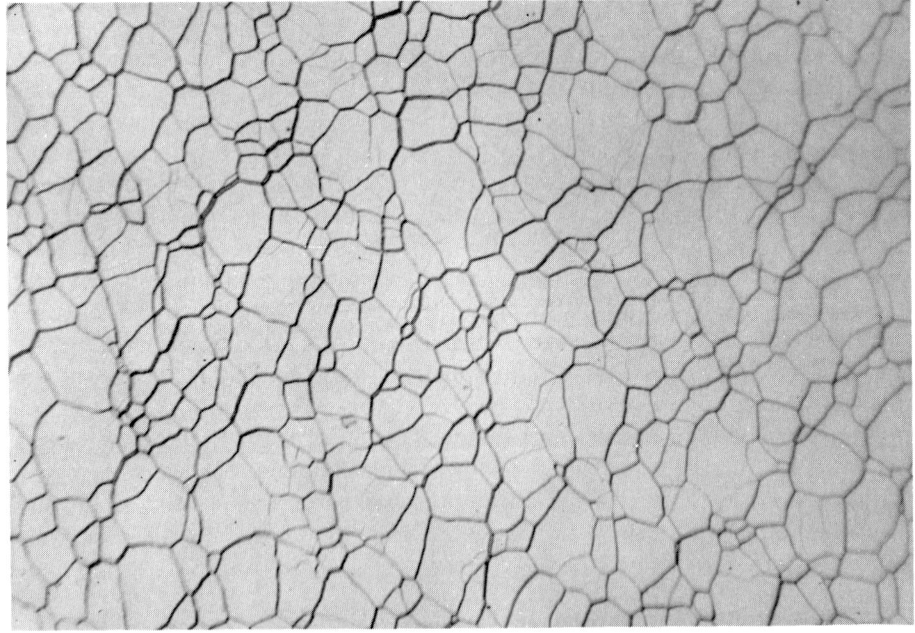

Figure 5.3 Grandjean planar texture of a cholesteric liquid crystal.

solutions, are described as optically active since the right- and left-handed components of plane polarized light propagate at different velocities, causing the plane of polarization of the emergent beam to be rotated[8]. This property is often referred to as circular birefringence and gives rise to rotations of about 10°/mm. In the case of cholesteric liquid crystals (ChLCs), there is therefore a twofold optical activity: (1) molecular and (2) macromolecular — associated with the helical structure. When plane polarized light propagates through this structure its right-handed and left-handed components see different refractive indices, resulting in a phase retardation between them; therefore the resultant plane of polarized light is rotated. The important difference in this case is that the rotation is much greater, of the order of 10^{3}°/mm.

A further unique feature of the helical structure occurs when its optical wavelength (np) is equal to that of the incident light:

$$\lambda = np$$

where

n = refractive index

p = pitch length

In this case, circularly polarized light with the same sense as the helix is selectively reflected in a manner which is analogous to Bragg X-ray reflection. In order to visualize how this occurs, it is convenient to consider the helical structure as depicted in Figure 5.1. The arrows in this model represent the director orientation in each N-like layer. Therefore for each layer or sheet, two refractive indices may be defined: n_e parallel to the director and n_o at right angles to it in the same plane. If a plane polarized beam with the same wavelength as the helix enters the structure with its plane of polarization parallel to the director, and if the helix is right-handed, then one circularly polarized component will propagate seeing some average refractive index and be transmitted. However, the other component will see a sinusoidal variation in refractive index along the helix axis with n_e repeated at every 180° rotation or $p/2$ spacing and be scattered. The scattered light from each $p/2$ plane constructively interferes and is reflected as right-handed circularly polarized light, whose wavelength is given by

$$\lambda - n_e p$$

By the same process, incident plane polarized light with its plane orthogonal to the director will have 50 per cent of its intensity reflected and its wavelength is given by

$$\lambda' = n_o p$$

When ordinary white light is incident on the helix, all wavelengths between λ and λ' are reflected so that the bandwidth is given by

$$\Delta\lambda = p(n_e - n_o)$$

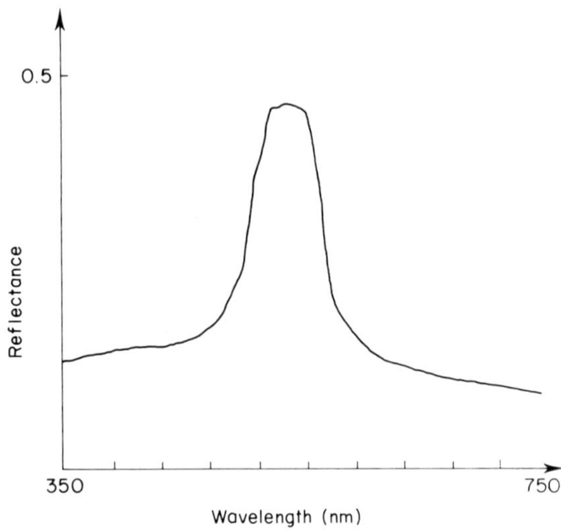

Figure 5.4 Reflectance spectrum of a cholesteric liquid crystal aligned in a 10 μm thick glass cell at normal incidence.

where

$$n = n_e - n_o$$

A typical reflectance spectrum is shown in Figure 5.4. As a consequence of this circular dichroism, ChLCs show an anomalous dispersion in their rotary power at the selective reflection wavelength (Figure 5.5); this is analogous to the Cotton effect[9] given by optically active isotropic liquids containing a chromophore.

Although in the above description of selective reflection an analogy was drawn to Bragg reflection[10], there are two important features which are different and are not adequately or easily explained by this simple model:

1. Only first-order reflections occur at normal incidence.
2. The reflected wavelength dependence on the angle of incidence does not obey Bragg's law:

$$\lambda = n\,2d \sin \theta$$

where

n = order

d = layer spacing

To explain these points accurately, more detailed theoretical arguments based on Maxwell's equation have been proposed by several authors[11].

The pitch length of the Ch helical structure normally shows an inverse temperature dependence with a slope of typically −0.3%/°C. However, this can

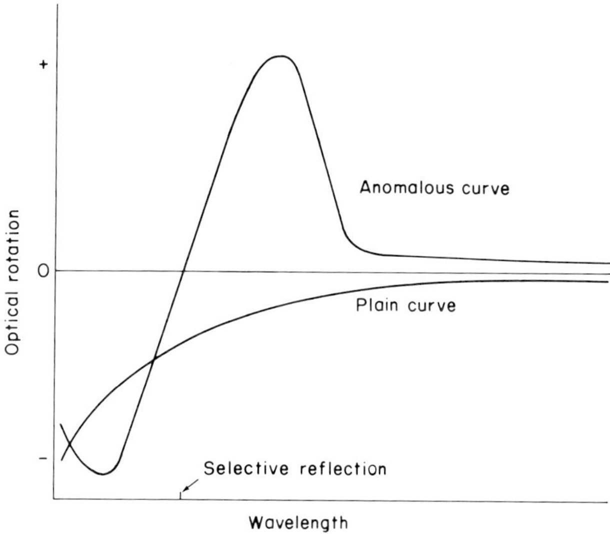

Figure 5.5 Anomalous optical rotary dispersion of a cholesteric liquid crystal at the selective reflection wavelength.

rapidly increase in some materials by several orders of magnitude as a result of the buildup of short-range S order just above an S–Ch transition. When the selective reflections are in the visible range, this rapid pretransitional unwinding of the pitch is seen as a dramatic colour change from blue to red on cooling.

The selective reflection and thermochromic properties of the Ch phase are undoubtedly very interesting phenomena and have stimulated imaginative ideas for applications[12]. Indeed, over the past twenty years, many hundreds of patents and papers have been published describing a remarkable range of thermometric, thermographic and other devices. However, despite the proliferation of potential uses, few have achieved commercial reality, and even these have received only limited market acceptance. This poses the question of whether this is due to an inadequacy in the ideas or in the technology. It is therefore worth while considering the applications before reviewing the materials and the developments in device construction.

5.3 Applications

Although within the scope of this chapter emphasis is placed on the selective reflection properties of short pitch length ChLCs, it is useful to remember that the materials are also of interest in electrooptical displays as dopants: (1) in twisted nematic displays[13] where they are used in concentrations of < 1 wt % to remove reverse twist defects; (2) in cholesteric to nematic dyed phase change

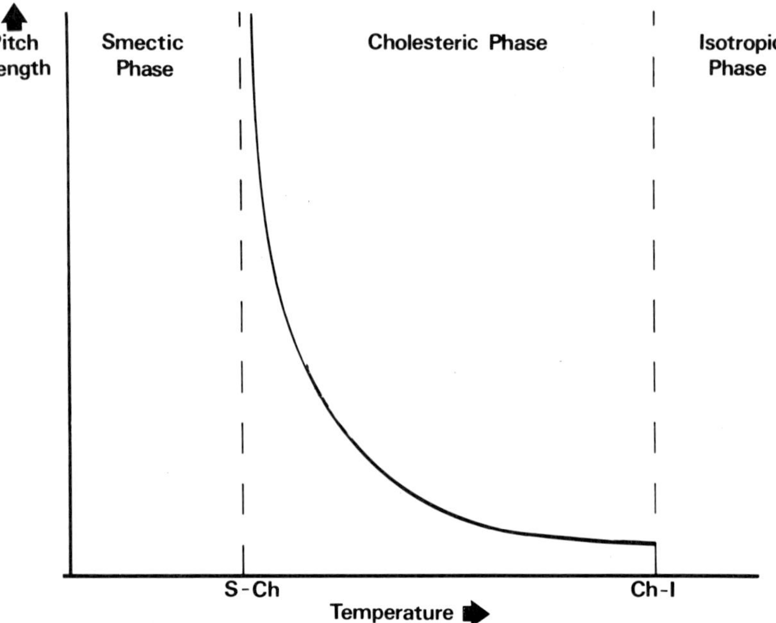

Figure 5.6 Temperature dependence of cholesteric helical pitch (and colour play) in the region of a smectic A transition.

displays[14] where they are used in concentrations of $\leqslant$ 10 wt % to enhance contrast by preventing wave-guiding. In recent years, these display applications have in fact stimulated a great deal of development work on Ch materials.

The thermochromic applications invariably use thin-film preparations of the cholesterogen which are viewed against a black background. Essentially these temperature-sensing devices may be placed into the following categories of applications.

5.3.1 Thermometry

In terms of volume, digital thermometers[15] are the most important products currently produced. A particular temperature is indicated by a green selective reflection occurring beneath a fixed legend. A multiplicity of legends, each associated with an appropriately formulated mixture, enables a variety of thermometers to be made. A selection of those commercially available is shown in Figure 5.7. These provide a simple, easy to read means of indicating temperature in the home, office, aquarium, etc. A particularly attractive device is the forehead thermometer, which, if accurately calibrated, can give a good indication of core body temperature in sleeping children[16] and postoperative patients still under anaesthetic[17].

Figure 5.7 Commercially available thermochromic liquid crystal devices.

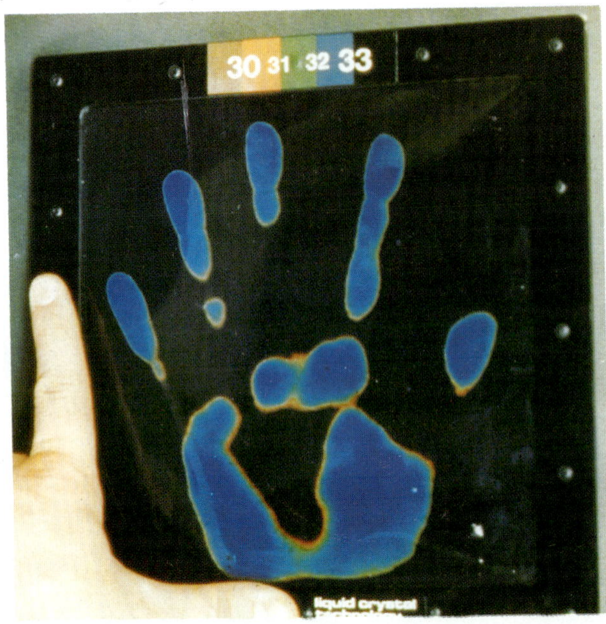

Figure 5.9 Thermograph of a human hand.

Figure 5.10 Thermograph of a printed circuit board.

In addition to using the rapid colour changes which occur near an S phase, ChLCs may be arranged in thin-film structures to make maximum thermometers by taking advantage of the irreversible texture change which occurs at the Ch to isotropic transition (see Figure 5.8). On heating the Grandjean selectively reflecting texture into the isotropic phase, the colour is lost, and upon cooling, the thermodynamically preferred, opaque, focal-conic texture forms. This property may be exploited in freeze/thaw indicators or in clinical thermometers[18]. Recharging may be effected by mechanically realigning the sample.

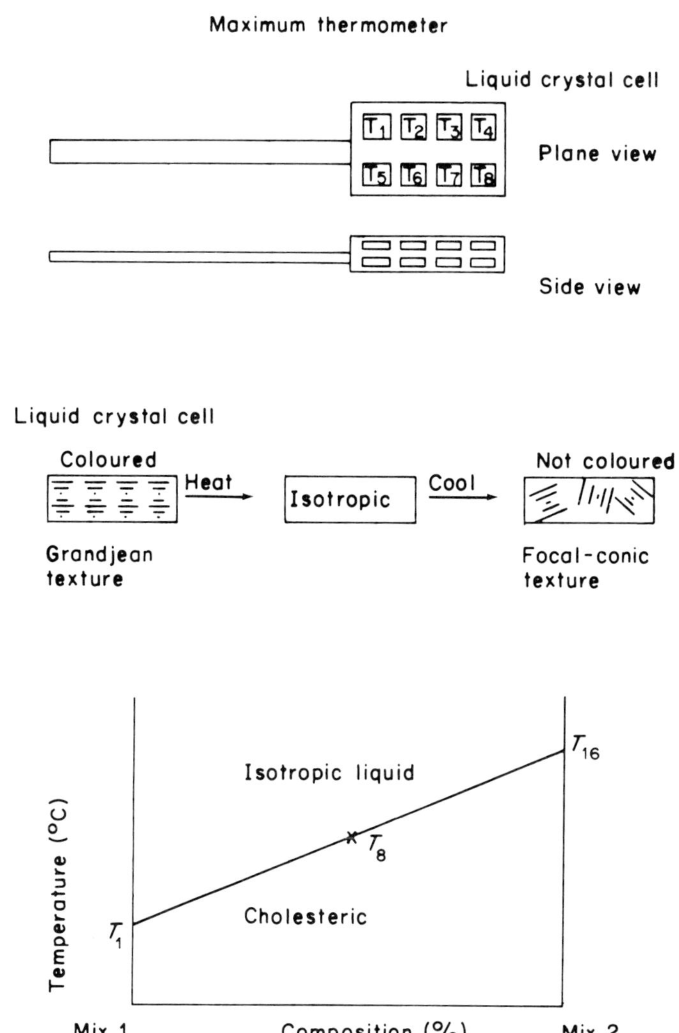

Figure 5.8 Clinical thermometer using the irreversible texture change at a Ch–I transition.

Memory devices[19] using ChLCs can also be obtained by rapidly quenching the coloured Grandjean state through the S phase and into a glassy state, before the helical structure can relax. Heating from the glassy state into the S state then causes the colour to be lost.

5.3.2 Medical thermography

Biomedical thermography[20] is an important diagnostic aid which is used to assist in the identification of a wide range of medical conditions. In many of these cases, some of which are given below by way of example, ChLCs have been evaluated as one means of producing visual thermal maps (Figure 5.9):

1. Breast cancer detection[21]
2. Placental location[22]
3. Vascular disorders[23]
4. Skin grafting[24]

Although the full list of applications is impressive, the reader is cautioned that (1) in many cases thermographic diagnosis is not wholly reliable and (2) there are inherent disadvantages associated with ChLCs relative to other non-invasive diagnostic techniques, e.g. infrared detectors. The disadvantage with ChLCs is that to ensure good thermal contact and visualization, either they must be painted on top of a black stain (unattractive to patients) or be stretched in a black elastomeric film over the area under investigation. In certain cases, the applied pressure of the films may lead to inaccuracies.

5.3.3 Non-destructive testing[25]

Thermal mapping of inanimate objects can provide (Figure 5.10) engineers with useful diagnostic and design information:

1. Detection of structural flaws in, for example, welds[26]
2. Hot spots in electrical circuits indicating shorts[27]
3. Heat transfer effects in aerodynamic models[28]

5.3.4 Radiation sensing

In general, these devices consist of a low thermal mass transducer which converts the radiation into heat energy, and the temperature rise is viewed directly as an induced colour change in a film of ChLC coating the surface. The technique has been investigated with both infrared[29] (thermal viewers, lasers) and microwave[30] (oven radiation leakage detectors) sources.

It is important to note that the thermal imaging applications are severely limited by thermal spread in the LC film. However, where detailed imaging is not essential, these direct-view transducers can provide a simple, low-cost, practical means of detecting radiation at power densities of a few milliwatts per square centimetre[31].

5.3.5 Decorative/novelty

Under this heading are included a vast category of uses in which the visual appeal of colour change with temperature is of primary importance. A particularly good example of this type of application is provided by Makow[32] who exploits the temperature and angle dependence of the colours of Ch paints to add extra dimensions to paintings. A number of fashion ideas have also been explored, including slogans on fabrics[33] and 'mood'-indicating jewellery[34].

So far all of the applications described exploit the temperature dependence of the selective reflection in the region of the S–Ch transition. However, this transition itself can be raised by the action of pressure[35] or lowered by the addition of impurities, with the effect of producing colour changes at a constant temperature. The effect of impurities on the selective reflection wavelength has stimulated interest in the use of cholesterogens in atmospheric pollutant detection[36].

Cholesterogens that are not strongly thermochromic have been studied as reflectors in twisted nematic displays[37]. The effects of electric fields have also been investigated and limited field tunable circular dichroism demonstrated[38]. Another imaginative idea from Makow[39] relates to the use of two layers of ChLC on either side of a half-wave plate with each layer portraying half of a steroscopic scene. The layers can then be viewed through spectacles, one element of which is a right-handed circular polarizing filter and the other a left-handed filter, to give a three-dimensional image.

5.4 Cholesteric materials

The first observation[40] of the Ch mesophase was made nearly one hundred years ago for an ester of cholesterol (Figure 5.11). Since then several hundred cholesterogens based on this naturally occurring optically active alcohol or other sterols have been synthesized[12].

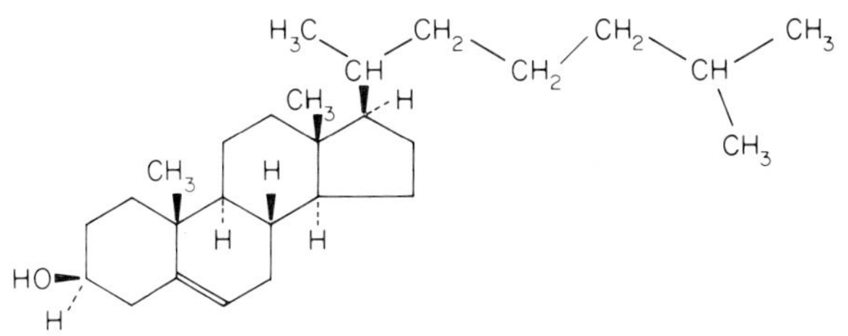

Figure 5.11 Structure of cholesterol.

Although the majority of this work was encouraged in the early 1960s by the potential use of the thermochromic properties, the practical use of sterol esters is limited by their chemical/photochemical instability, as well as by their limited range of physical properties. Despite these problems it is surprising that cholesteryl esters still feature prominently in many applications. Indeed the shortcomings of these materials would appear to have biased the commercial exploitation into novelty markets in which accurate colour/temperature calibration is not essential.

This situation is in marked contrast to the rapid development of NLCs for the rapidly expanding electrooptical display industry. However, the need for cholesterogens in the display technologies, as described above, has provided spin-off materials developments which are now stimulating renewed interest in thermochromic devices.

The chemical key to the synthesis of non-sterol cholesterogens is the availability of *S*-(−)-2-methylbutan-1-ol in high optical purity. Using synthetic procedures which preserve the optical activity, this naturally occurring alcohol has been incorporated as a chiral alkyl group into the core structures of many known NLCs. This work has provided a wide range of new Ch materials, as shown in Table 5.1.

These have enabled a more systematic study to be made of molecular structural effects on pitch length and helical twist sense. Moreover, the new chiral nematics offer advantages of:

1. Greater stability by the selection of chemically and photochemically stable functional groups.
2. Wider range of thermodynamic and other physical properties such as birefringence through the selection of a variety of different core structures.
3. Colour play and colour/temperature linearity control by dilution of optical purity with racemic material.

The LC behaviour associated with different core structures of chiral nematics is the same as for nematics, which has already been dealt with in Chapter 2. Therefore, using selected examples, emphasis is placed in this section on examining the effects of the chiral alkyl group on the pitch length and twist sense.

The chiral 4-alkyl-4′-cyanobiphenyls[41] were among the first chiral nematics to be synthesized and a number of important points concerning the position of the asymmetric centre in relation to the aromatic core of the molecule were thereby established (Table 5.2):

1. Chain branching gives rise to an expected lowering of the transition temperatures.
2. The closer the asymmetric centre is to the rigid core of the molecule the shorter is the pitch length. Note, however, the exceptional, long pitch for compound II, $n = 1$, compared with compound II, $n = 3$, in Table 5.2.
3. Chiral alkyl groups give much shorter pitch lengths than chiral alkyloxy groups.
4. When the chiral centre has an S absolute configuration and is at an *even* number of atoms from the rigid core, then the optical twist sense of the Ch helices is *dextro*, and, conversely, when at an *odd* number it is *laevo*.

Table 5.1 Typical chiral nematic liquid crystals

Structure	References
R—⌬—⌬—CN	41, 42
R—⌬—⌬—⌬—CN	41, 42
R—⌬—⌬—CH_2CH_2—⌬—CN	42
R—⌬—OCO—⌬—R′ [a]	43, 44, 45
R—⌬—OCO—(naphthalene)—R′	43
R—⌬—OCO—⌬—⌬—R′	43, 45
R—⬡—OCO—⌬—R′	43
R—⌬—OCO—⌬(X)—R′ X = H, F	46
R—⌬—B—⌬—R′ B = N=N	47
N=CH	48
C≡C	49

[a]Ester links have been depicted as —OCO— to indicate that the linkage may be both —COO— and —OOC—.

As the enantiomers would be expected to give the opposite optical twist sense[4], these results and those reported by others[50] confirm the applicability of the empirical rule[5]:

Right-handed optical twist sense of helices	Left-handed optical twist sense of helices
SED	SOL
ROD	REL

S and R refer to absolute configuration.

D and L refer to dextro and laevo optical twist senses of the helix.

E and O refer to an even or odd position of the chiral centre from the rigid molecular core.

Table 5.2 Properties of chiral nematic biphenyls

$CH_3CH_2\overset{*}{C}H(CH_3)(CH_2)_n$–⟨biphenyl⟩–CN

I

$CH_3CH_2CH(CH_3)(CH_2)_nO$–⟨biphenyl⟩–CN

II

Structure	n	Melting point (°C)	Ch–I (°C)	S_A–Ch (°C)	Pitch (μm)	Optical twist sense of helix
I	1	4	(−30)	(<−50)	0.15	D
I	2	9	(−14)	(−22)	0.3	L
I	3	28	(−10)	(−20)	0.4	D
II	1	54	(9)	—	1.5[a]	L
II	3	58	(35)	—	1.0	L

D = right-handed.
L = left-handed.
( ) Monotropic transition.
[a]See text.

As emphasized earlier, most of the work on structural optical twist sense relations has been carried out with cholesterogens containing chiral alkyl groups, and the above rules relate specifically to such systems. More recently[51], it has been shown that if the substituent on the chiral centre has a negative inductive effect, e.g. 2-chloro- or 2-cyano-propyl, the rule is changed to SEL, SOD and ROL, RED.

The results above and those in Table 5.2 demonstrate the effect of moving the chiral centre along the alkyl chain on (1) the Ch helical pitch and (2) the optical twist sense, which alternates successively between left- and right-handed. The two ring esters in Table 5.3 illustrate the effect of attaching the chiral group to different ring systems.

In each case the *S*-(−)-2-methylbutyl group is attached directly to the ring and therefore all the Ch helices are right-handed in optical twist sense. The first two examples in Table 5.3 show that the pitch length is unaffected by whether the chiral group is in the aromatic acid or the phenol moiety. However, depending on whether the chiral group is attached to a benzene, cyclohexane or bicyclo(2.2.2)octane ring, its twisting power can change by a factor of two.

This gives the following order of efficiency for twisting power:

–⟨benzene⟩– > –⟨bicyclo(2.2.2)octane⟩– > –⟨cyclohexane⟩–

The addition of two chiral groups to a molecule can either show an additive or subtractive effect in the twisting power depending on whether they are of the same or different handedness, as shown by the compounds in Table 5.4.

Table 5.3 Properties of two-ring esters

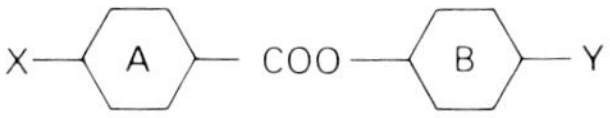

X—(A)—COO—(B)—Y

X	A	B	Y	Melting point (°C)	Ch–I (°C)	Pitch (μm)	Optical twist sense of helix	References
2MB	Ph	Ph	OC_6H_{13}	22	(17)	0.23	D	43
$C_6H_{13}O$	Ph	Ph	2MB	38	(36.7)	0.23	D	43
2MB	Cy	Ph	OC_6H_{13}	50	(29.5)	0.43	D	43
C_3H_7	Cy	Ph	2MB	26.5	(1)	0.23	D	43
2MB	BCO	Ph	OC_6H_{13}	43	50	0.36	D	46
C_6H_{13}	BCO	Ph	2MB	49.5	(36.5)	0.23	D	46

D = right-handed
( ) Monotropic transition.
2MB = $CH_3CH_2{}^*CH(CH_3)CH_2$—

Ph =

Cy =  (*trans*)

BCO =

Table 5.4 Properties of esters containing two chiral groups

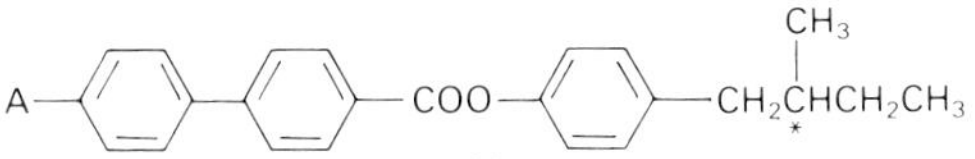

A—Ph—Ph—COO—Ph—$CH_2\overset{*}{C}H(CH_3)CH_2CH_3$

IV

A	Melting point (°C)	S_A–Ch (°C)	Ch–I (°C)	Pitch (°C)	Optical twist sense of helix	References
2MB	103	—	115.5	0.10	D	43
2MBO	88	124	145	0.46	D	5

D = Right-handed
2MB = $CH_3CH_2{}^*CH(CH_3)CH_2$— : SED
2MBO = $CH_3CH_2{}^*CH(CH_3)CH_2O$— : SOL

The first material has two SED chiral centres which add to give a pitch length of 0.1 μm. However, the second material has one SOL and one SED, and in this case, although the SED group dominates, giving a resultant right-handed optical twist sense of the helix, the overall twisting power of the molecule is attenuated by the presence of the SOL group.

The qualitative rules outlined above provide two useful functions:

1. They assist in the design of new materials.
2. They provide guidelines for the formulation of thermochromic and other Ch mixtures.

However, it is important to repeat that the observations referred to in Tables 5.3 and 5.4 all relate to cholesterogens derived from the same optically active alcohol i.e. *S*-(−)-2-methylbutan-1-ol.

The generality of all these qualitative relationships between molecular structure, twisting power and helical twist sense requires testing by the examination of alternative optically active starting materials, in which the absolute configurations are accurately assigned. In this context a study of new chiral SLCs[52, 53] (see Chapter 4) should add further to our understanding. In addition to using chiral alkyl and substituted alkyl groups, it is useful to note that optically active LC materials can also be made from molecules with axial chirality, e.g. spiro compounds[54, 55].

5.5 Thermochromic mixtures

Commercial thermochromic applications require materials with low melting points, short pitch lengths and smectic transitions just below the required temperature-sensing region. Satisfaction of these requirements over the desired operating temperature range (−50 to +100 °C) necessitates the formulation of mixtures. A convenient mixture 'system' has been described[56] and is available commercially (BDH Ltd, Poole, Dorset, UK). The system comprises four mixtures (Figure 5.12) formulated from homologues of esters (V–VII[43, 57] in Table 5.5). An attractive commercial feature in the molecular design of these esters is that they incorporate the same optically active phenol moiety, while enabling the range of desired thermodynamic properties to be obtained by varying the acid groups.

The 'A' mixtures, TM74 and TM75, are the low- and high-temperature formulations respectively and these are mutually miscible in all proportions. Therefore the temperature at which a desired colour play begins can be selected at any point between the lower and higher S_A–Ch transition temperatures of TM74 and TM75 respectively by mixing them according to the phase diagram shown in Figure 5.13.

The 'B' analogue mixtures are partially racemized and enable the temperature band of the colour play to be extended from ~0.5 to ~5 °C. Some typical variations in selective reflection wavelengths with temperature for these mixtures are shown in Figure 5.14. Alternative mixtures are also available[58] which have thermochromic properties up to 110 °C.

The formulation of thermochromic mixtures is complicated by the need to retain both low melting points and high S–Ch transition temperatures. In practice this limits the range of suitable components quite considerably and makes the esters (VI) of particular importance. In order to overcome this problem, injected S phase [59] behaviour has been investigated[60] in 'hybrid' mixtures of strongly polar and weakly polar chiral NLCs (Figure 5.15).

Materials for thermochromic applications

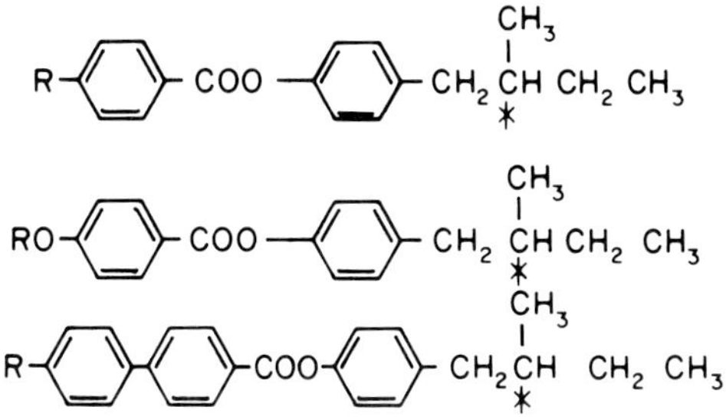

$R-C_6H_4-COO-C_6H_4-CH_2\overset{*}{C}H(CH_3)CH_2CH_3$

$RO-C_6H_4-COO-C_6H_4-CH_2\overset{*}{C}H(CH_3)CH_2CH_3$

$R-C_6H_4-C_6H_4-COO-C_6H_4-CH_2\overset{*}{C}H(CH_3)CH_2CH_3$

⁎ Denotes chiral centre

TM 74 A		TM 74 B†	
S_A–Ch	−32.6 °C	S_A–Ch	−32.3 °C
Ch–I	15.9 °C	Ch–I	16.2 °C

TM 75 A		TM 75 B†	
S_A–Ch	41.3 °C	S_A–Ch	42.4 °C
Ch–I	53.2 °C	Ch–I	54.4 °C

† B mixtures contain 50 wt % of racemic material.

Figure 5.12 Transition temperatures of BDH mixtures TM74 A/B and TM 75 A/B.

In summary, thermochromic mixtures with injected S phases demonstrate three important points:

1. The twisting power of the polar chiral N (CB15) is host dependent and is greater in non-polar hosts.
2. The availability of a wider range of materials for thermochromic mixtures enables greater control over other physical properties such as birefringence, which determines the width ($\Delta\lambda$) of the selective reflection band.
3. The extent of the pretransitional effects of the S phase varies across the phase diagram, providing another means of changing $d\lambda/dT$.

The thermochromic behaviour which occurs across the phase diagram in Figure 5.15 is shown in Figure 5.16.

Table 5.5 Properties of esters derived from *S*-4-(2-methylbutyl)phenol

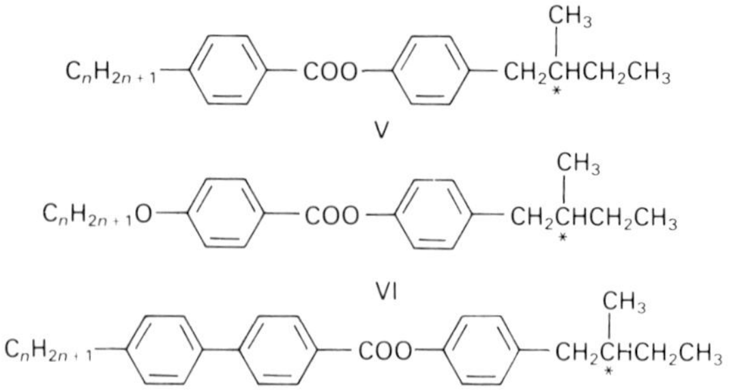

C_nH_{2n+1}—⟨ring⟩—COO—⟨ring⟩—$CH_2\overset{*}{C}H(CH_3)CH_2CH_3$

V

$C_nH_{2n+1}O$—⟨ring⟩—COO—⟨ring⟩—$CH_2\overset{*}{C}H(CH_3)CH_2CH_3$

VI

C_nH_{2n+1}—⟨ring⟩—⟨ring⟩—COO—⟨ring⟩—$CH_2\overset{*}{C}H(CH_3)CH_2CH_3$

VII

Structure	*n*	Melting point (°C)	S_A–Ch (°C)	Ch–I (°C)	Pitch (μm)	Optical twist sense of helix
V	3	17	—	—	0.23	D
V	5	9	—	—	0.23	D
V	7	16.5	[−32]	(−1.8)	0.23	D
V	9	22	(−6.0)	(8.0)	0.23	D
VI	6	38	(17)	(37)	0.23	D
VI	7	45	(15)	(36.3)	0.23	D
VI	8	47	(19)	(42)	0.23	D
VI	9	50	(37)	(43)	0.23	D
VI	10	42	42.2	45.3	0.23	D
VII[a]	7	80	139	148	0.23	D
VII[a]	9	50	136	139	0.23	D

D = right-handed.
( ) Monotropic transition.
[ ] Virtual transition.
[a]These also exhibit chiral S_C and other smectic phases[62].

5.6 Thermochromic device construction

In general, thermochromic LC devices have a thin film of cholesterogen sandwiched between a transparent supporting substrate and a black absorbing layer. The preferred fabrication method, which has evolved from the 1960s, involves producing an 'ink' with the LC by encapsulating it in a polymer and using printing technologies to apply it to the supporting substrate. Two methods of manufacturing the inks have been developed and are in current use. These are:

1. Gelatin microencapsulation[61]
2. Polymer dispersions[62, 63]

The first method encases droplets of LC (diameters 5–50 μm) in polymer shells made from one or a combination of (1) gelatin/gum acacia, (2) polyvinyl alcohol, (3) resorcinol formaldehyde. A slurry of the capsules in a water/polymer solution gives the ink. The second method is similar to the first, but differs in that the LC droplets are dispersed in a continuous polymer matrix without precoating with a polymer shell (Figure 5.17).

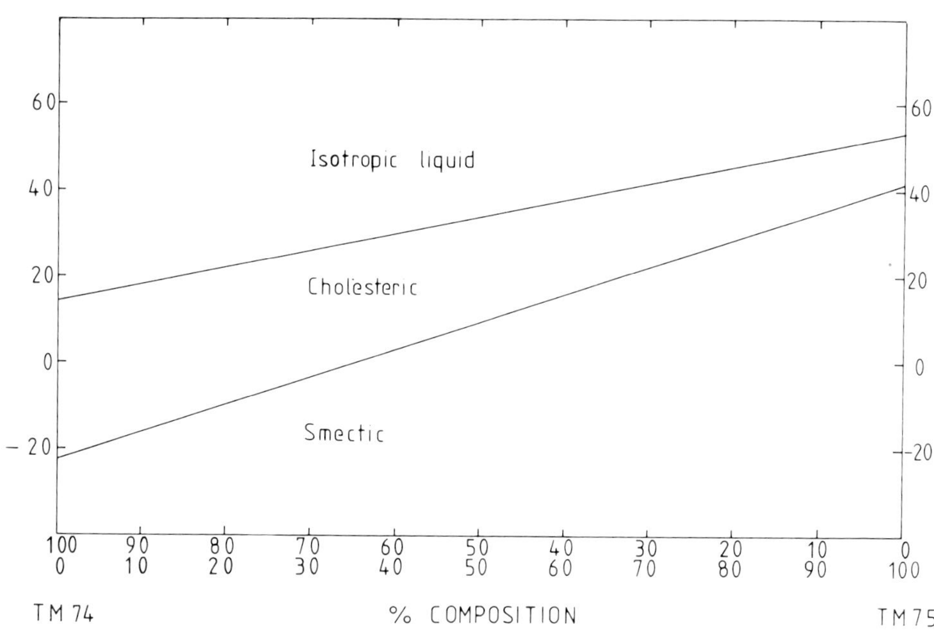

Figure 5.13 Phase diagram for TM74 and TM75.

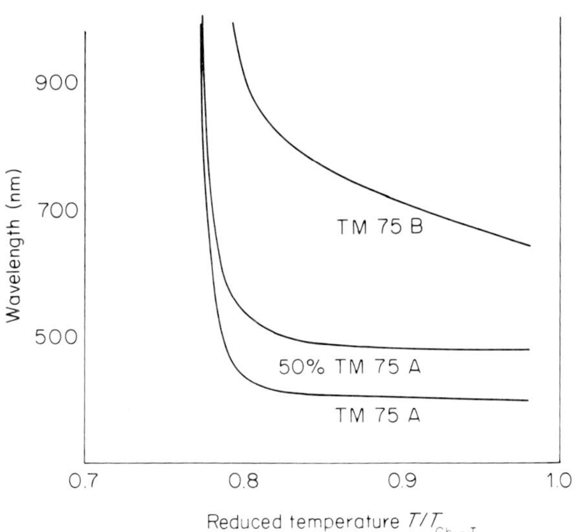

Figure 5.14 Wavelength of selective reflection as a function of reduced temperature for TM 75 A and B and a mixture.

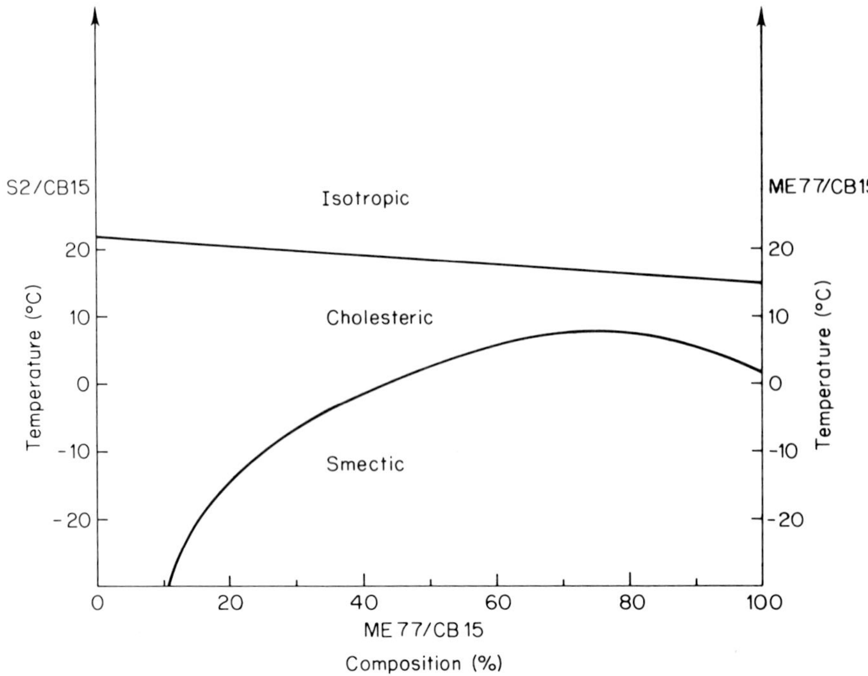

Figure 5.15 Phase diagram of a polar biphenyl liquid crystal mixture (S2/CB15) with a non-polar dialkylbenzoate ester mixture (ME77/CB15).

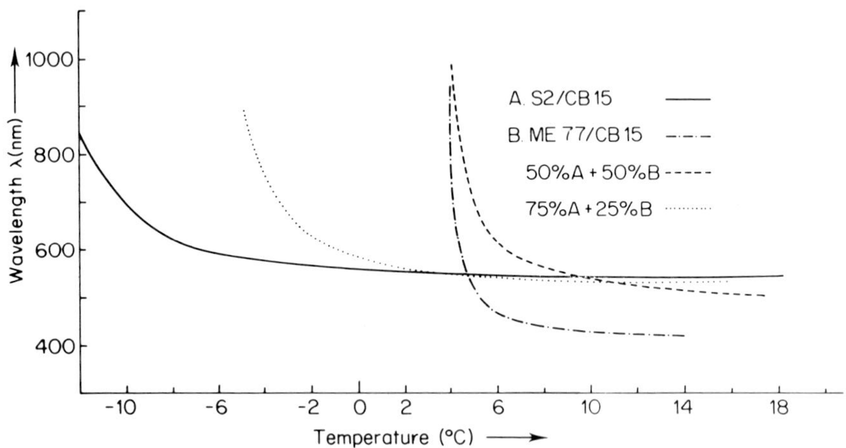

Figure 5.16 Wavelength of selective reflection as a function of temperature for some compositions of polar and non-polar cholesterogens (see also Figure 5.15).

The objectives of these encapsulation procedures are to support the LC in a dry film with the minimum degradative effect on the intensity and colour purity ($\Delta\lambda$) of the reflected light. However, some degradation in the optical appearance is inevitable since, for an optimum reflectance of 0.5, all the helices have to be aligned, and in encapsulated samples this is not possible. In practice the idealized spheres as shown in Figure 5.17 are likely to be flattened to some degree as the binder polymer contracts on drying. Indeed, evidence has been found[63] that the S–Ch transition temperature increases as the binder hardens, suggesting a pressure increase in the capsules. Although these distorted spheres are more favourable structures for formation of the Grandjean texture, large areas within the droplets adopt focal-conic textures and reflectances drop to ~0.1. In addition to the disadvantages of low brightness and calibration control, difficulties are often encountered in achieving compatibility between the binding polymer, solvent and the surface to be coated.

These problems have been overcome by a new technique[64] for preparing well-aligned thin-film structures of ChLCs, which involves laminating the LC between two embossed plastic sheets (Figure 5.18).

The embossed structures fulfil two functions:

1. The fine structure of microgroves aligns the LC.
2. The coarse structure provides spacing and containment walls.

Encapsulation techniques

(a) Microencapsulation dispersion

Cholesteric liquid crystal

Binder polymer, e.g. PVA

Capsule coating, e.g. gum arabic/gelatine

Substrate, e.g. mylar

(b) Polymer dispersion

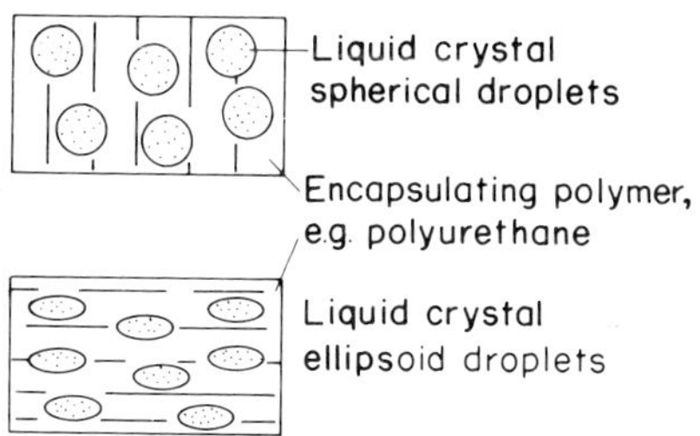

Figure 5.17 Idealized model of encapsulated liquid crystals.

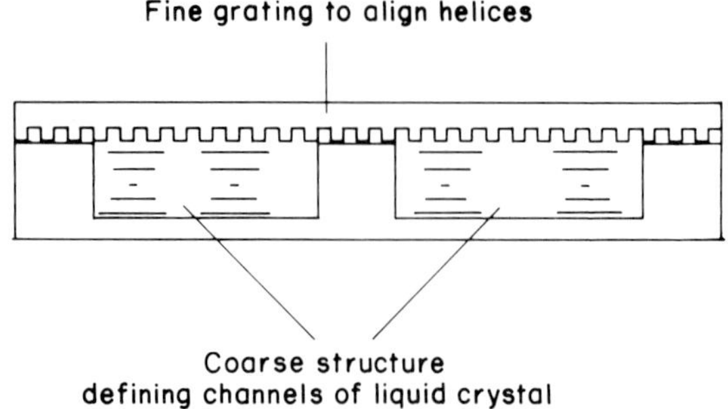

Figure 5.18 Embossed structures on laminating plastic films.

Selective reflection intensities from these devices are three times greater than those for microencapsulated devices[63].

A further point to note about these new structures is that the high degree of transmission of the films enables them to be used in bilayer arrangements with a $\lambda/2$ plate to double the reflectivity. The role of the $\lambda/2$ plate is to reverse the sense of the transmitted circularly polarized light from the first layer of LC so that it is reflected by the second layer. Thus both circularly polarized components can be obtained from two identical selectively reflecting Ch layers. The same effect can in principle be obtained by optically coupling a right-handed and left-handed cholesterogen, e.g. an SED with its chemically equivalent REL[65]. The efficiency of such structures could be of use in broad- and narrow-band filters as well as in dichroic mirrors.

5.7 Conclusions

In the last ten years, the development of chiral NLCs has made available a large number of chemically and photochemically stable cholesterogens with a wide range of physical properties. Qualitative relationships have been found for these materials between molecular structure, helical pitch length and twist sense. Our understanding of these relationships will be improved further by the new chiral LC materials being developed for chiral S_C devices.

Thermochromic mixtures using the new materials are commercially available and offer colour plays between 0.5 and 5 °C over the temperature range −30 to +110 °C. These developments are focusing attention on the inadequacies of brightness and contrast in thermochromic devices fabricated by encapsulation methods. New device structures have been developed to overcome these problems using embossed diffraction gratings to promote alignment of the liquid crystal.

The combination of these two developments should help the temperature-sensing application of cholesterogens to realize their true potential.

5.8 Acknowledgements

The author would like to express thanks to colleagues at RSRE, Malvern, BDH Ltd, Poole, Dorset, The University of Hull and Liquid Crystal Devices Ltd, Ruislip, for many helpful discussions.

5.9 References

1. C.W. Oseen, *Trans. Faraday Soc.*, **29**, 883 (1933).
2. J.C. Martin and R. Cano, *Nouv. Rev. Opt.*, **7**, 265 (1976).
3. H. Stegemeyer and K.J. Mainvsch, *Naturwissenschaften*, **58**, 599 (1971).
4. M. Leclercq, J. Billard and J. Jacques, *Mol. Cryst. Liq. Cryst.*, **8**, 367 (1969).
5. G.W. Gray and D.G. McDonnell, *Mol. Cryst. Liq. Cryst.*, **34**, 211 (1977).
6. Y. Bouligand, *J. Phys. (Paris)*, **34**, 603 (1973).
7. P.P. Crooker, *Mol. Cryst. Liq. Cryst.*, **98**, 31 (1983).
8. G. Hallas, *Organic Stereochemistry*, McGraw-Hill, London (1965).
9. H. Stegemeyer and K.J. Mainusch, *Chem. Phys. Lett.*, **6**, 5 (1970).
10. J.L. Fergason, *Mol. Cryst. Liq. Cryst.*, **1**, 293 (1966).
11. D.W. Berreman and T.J. Scheffer, *Phys. Rev. Lett.*, **25**, 557 (1970); R. Dreher and G. Meier, *Phys. Rev. A*, **8**, 1616 (1973); A. Saupe and G. Meier, *Phys. Rev. A*, **27**, 2196 (1983).
12. G.H. Brown (ed.), *Advances in Liquid Crystals*, Vol. 2, Academic Press, New York (1976), pp. 73–172.
13. E.P. Raynes, *Electron. Lett.*, **9**, 118 (1973).
14. D.L. White and G.N. Taylor, *J. Appl. Phys.*, **45**, 4718 (1974).
15. R. Parker, US Pat. 3,822,594 (1974).
16. D.A. Swanklin, 'Problem definition study on liquid crystal forehead temperature strips', prepared by the National Consumers' League, US (1981).
17. D. Lees, W. Schuette, J. Bull, J. Whang-Peng, E. Atkinson and T. Macnamara, *Anesth. Anal.*, **57**, 669 (1978).
18. C. Hilsum and D.G. McDonnell, Br. Pat. 2,085,585B (1985).
19. Stafreez (Trademark) Biosynergy Inc., 724 West Algonquin Road, Arlington Heights, IL 60005, USA (1983).
20. I. Nyirjesy, M.R. Abernathy, F.S. Billingsley and P. Bruns. *J. Reproductive Med.*, **18**(4), 165 (1977).
21. M. Gautherie and C.M. Gros, *Cancer*, **45**, 51 (1980).
22. R.C. Margolis and L.S. Shaffer, *J. Am. Obstetrics Assn.*, **73**, 910 (1974).
23. C. Ambrosi and C. Bourcle, *Gaz. Med. France*, **82**, 628 (1975).
24. B.Y. Lee, S.S. Trainov and J.L. Madden, *Arch. Phys. Med. Rehab.*, **54**, (1973).
25. G.D. Dixon, *Mater. Eval.*, **35**, 51 (1977).
26. W.E. Woodmansee and H.L. Southworth, *Mater. Eval.*, **26**, 149 (1968).
27. P.L. Garbarino and R.D. Sandison, *J. Electrochem. Soc.*, **120**, 834 (1973).
28. P. Ireland and T.V. Jones, 'Heat transfer and cooling in gas turbines', AGARD Conf. Proc. CP390 (1985).
29. R.D. Ennulat and J.L. Fergason, *Mol. Cryst. Liq. Cryst.*, **13**, 149 (1971); L.M. Klyukin, A.S. Sonin, B.M. Stepanov and I.N. Shibaev, *Kvantovaya Elektron.*, **2**, 61 (1975).
30. A.V. Tolmachev, E.Y. Govorun and V.M. Kuzmichev, *Zh. Eksp. Teor. Fiz.*, **63**, 583 (1972).
31. R.G. Pothier, US Pat. 3,713,156 (1970).
32. D. Makow, *Colour Res. Applic.*, **4**(1), 25 (1979).
33. A. Mace, J. Bersan and J. Luby, Fr. Pat. 2,461,008 (1981); K. Kurosawa, Jpn. Pat. 81,118,859 (1981).
34. B.G. James, US Pat. 3,802,945 (1974).
35. R. Shashidhar and S. Chandrasekhar, *J. Phys. (Paris)*, **36**, pE1–49, No. 3 (1975); P. Pollmann and H. Stegemeyer, *Chem. Phys. Lett.*, **20**, 87 (1973); R. Merran, Fr. Pat. 2,105,524 (1970).
36. T.K. Koka, Jpn. Pat. 850,525 (1985); J.L. Fergason, *Sci. American*, **211**, 77 (1964).
37. T.J. Scheffer, *J. Phys. D, Appl. Phys.*, **8**, 1441 (1975).
38. I. Fedak, R. Pringle and G.H. Curtis, *Mol. Cryst. Liq. Cryst.*, **64**, 69 (1980).

39. D. Makow, *Mol. Cryst. Liq. Cryst.*, **99**, 117 (1983).
40. F. Reinitzer, *Monatsh. Chem.*, **9**, 421 (1888).
41. G.W. Gray and D.G. McDonnell, *Electron. Lett.*, **11**, 556 (1975).
42. G.W. Gray and D.G. McDonnell, *Mol. Cryst. Liq. Cryst.*, **37**, 189 (1976).
43. G.W. Gray and D.G. McDonnell, *Mol. Cryst. Liq. Cryst.*, **48**, 37 (1978).
44. Y.Y. Hsu and D. Dolphin, *Mol. Cryst. Liq. Cryst.*, **42**, 319 (1977).
45. B.H. Klanderman and T.R. Criswell, *J. Am. Chem. Soc.*, **97**, 1585 (1975).
46. G.W. Gray and S.M. Kelly, *Mol. Cryst. Liq. Cryst.*, **95**, 101 (1983).
47. Y. Marakami, S. Furuyama and K. Morimoto, Jpn. Pat. 76,118,734 (1976).
48. Y.Y. Hsu and D. Dolphin, *Mol. Cryst. Liq. Cryst.*, **42**, 327 (1977).
49. I. Otkrytiya, Russ. Pat. application 754,815; Priority No. 2,713,095 (1979).
50. G. Heppke and F. Oestreicher, *Z. Naturforsch.*, **32A**, 899 (1977).
51. J.W. Goodby, *Science*, **231**, 350 (1986).
52. J.W. Goodby and T.M. Leslie in *Liquid Crystals and Ordered Fluids*, Vol. 4, ed. by J.F. Johnson and A.C. Griffin, Plenum Press, New York (1982), p. 1.
53. R. Blinc, N.A. Clark, J. Goodby, S.A. Pikin and K. Yoshino, (eds.), *Ferroelectrics*, Vols 58 and 59, Gordon and Breach, New York (1984).
54. F.P. Shvartsman and V. Krongauz, *J. Phys. Chem.*, **88**, 6448 (1984).
55. F.P. Shvartsman, I.R. Cabrera, A.L. Weis, E.J. Wachtel and V.A. Krongauz, *J. Phys. Chem.*, **89**, 3941 (1985).
56. D.G. McDonnell, 'Thermochromic liquid crystals', Information Booklet, BDH Ltd, Poole, Dorset (1980).
57. D.G. McDonnell, PhD thesis, The University of Hull (1979).
58. 'Thermochromic mixtures', Product Data Sheets, BDH Ltd, Poole, Dorset.
59. M.J. Bradshaw and E.P. Raynes, *Mol. Cryst. Liq. Cryst.*, **91**, 145 (1983).
60. J. Constant and D.G. McDonnell, Br. Pat. 85/15510.
61. US Pat. 3,585,318 (1969).
62. US Pats. 1,161,039 (1968) and 3,872,050 (1973).
63. P. Bonnett and D.G. McDonnell, unpublished results presented at the 11th International Liquid Crystal Conference, Berkeley, California (1986).
64. C. Hilsum, K.E.N. Kerr and D.G. McDonnell, Br. Pat. 2,143,323A (1985).
65. D. Makow, *Appl. Opt.*, **19**, 1274 (1980).

6 Liquid crystal polymers

H. Finkelmann
Institut für Makromolekulare Chemie
Universität Freiburg

6.1 Introduction

The liquid crystalline (LC) state, located between the crystalline and the isotropic liquid states of matter, offers broad perspectives for the application of materials that exhibit anisotropic but liquid properties. Electrooptical devices and thermography, mentioned in previous chapters, are two important examples which depend not only on the anisotropic physical properties but also on the liquid state of the LC material.

The LC state of matter is closely related to the molecular structure of the material, as described in Chapter 2. The basic molecular feature is a more or less rigid anisometric molecular shape, i.e. a cylindrical lath-like or disc-like molecular shape is required for the appearance of the LC state. Thousands of substances have been synthesized during the past few decades following this principle[1]. Because of their direct applicability in technology, low molar mass LC substances have been intensively considered, whereas related aspects for macromolecular LC compounds are still in their infancy.

Conventional amorphous, crystalline and partially crystalline synthetic polymers have become highly important materials, used in nearly all technologically important areas. The development of new polymers during the last decade has led to high-performance materials which offer better properties than conventional materials. Reinforced composites for modern aircraft, polymer alloys for automobiles, photoresists for the production of integrated circuits for the computer industry and polymers for controlled drug release are just some examples.

A new aspect in polymer science is the realization of LC polymers. These polymers combine the polymer specific properties (e.g. their ease of processing) with the anisotropic properties of the LC state. What then are the possibilities for a chemist to synthesize polymers exhibiting the LC state?

As mentioned above, the condition for the appearance of the LC state is a rigid, anisometric molecular shape. This concept can be directly applied to polymers, the monomer units (m.u.) of the polymer backbone containing the anisometric (mesogenic) moieties. If we restrict ourselves to the most important LC materials having a cylindrical or lath-like molecular shape, two basically different types of polymers are conceivable (Figure 6.1).

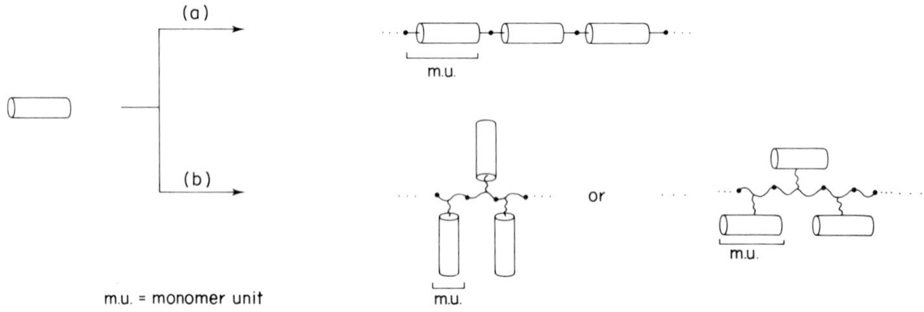

Figure 6.1 Schematic representation of the chemical constitution of LC polymers: (a) 'LC main chain polymers', (b) 'LC side chain polymers'.

(a) The mesogenic moieties are connected head to tail, forming the polymer main chain. Polymers having this chemical constitution are called 'LC main chain polymers'.
(b) The mesogenic moieties are attached as side chains to the monomer units of the polymer backbone. These polymers are called 'LC side chain polymers'.

In addition to the synthesis of homopolymers, which have only one type of monomer unit along the polymer main chain, a variety of copolymers have been prepared. These copolymers have different mesogenic moieties, including monomer units of both types (a) and (b), or non-mesogenic monomer units together with mesogenic monomer units. Furthermore, the linear primary structure of the main chain can be varied to yield branched or crosslinked main chains. Crosslinked polymers (e.g. elastomers) are of special interest, because of their elastic properties.

Following the classification mentioned above, some basic aspects of the synthesis of polymers exhibiting the LC state will be described in this chapter. Their characteristic properties will also be described and compared with the properties of conventional low molar mass LCs (LLCs). Aspects of applications will also be mentioned.

6.2 Liquid crystal main chain polymers

Ishihara and Onsager[2] and Flory[3], using a lattice model, predicted that a solution of rigid, rod-like macromolecules at a certain concentration would separate into two phases, one being anisotropic. This was experimentally confirmed in 1956 for a polypeptide which formed a stable, rod-like helical conformation in solution[4].

The systematic synthesis of rigid, rod-like polymers did not, however, begin until the mid-1970s, when it was observed that fibres spun from anisotropic solutions or melts exhibited extraordinary mechanical properties, such as high tensile strength and bending moduli[5–7].

For the preparation of LC main chain polymers, different models can be considered; these are shown schematically in Figure 6.2.

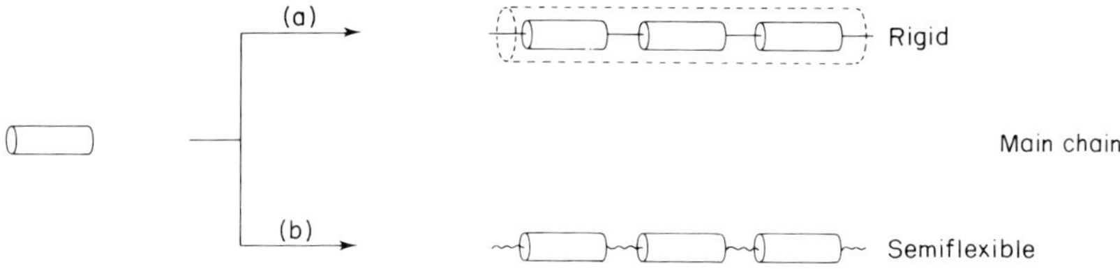

Figure 6.2 Schematic structure of LC main chain polymers. (a) Mesogenic moieties are directly connected, forming a rigid, rod-like polymer main chain. (b) Mesogenic moieties are connected via a flexible spacer, forming a semiflexible polymer main chain.

The mesogenic monomer units can be directly connected within the polymer main chain (Figure 6.2a). The resulting rigid, rod-like polymer can be described by the molecular axial ratio, which is directly proportional to the degree of polymerization (X). If the mesogenic groups are connected via a flexible spacer (Figure 6.2b), the main chain becomes semiflexible. Assuming a sufficient spacer length, the backbone can be described as a chain of freely-joined rods which have the axial ratio of the original mesogenic unit. These two different chemical structures for the polymers yield different phase behaviours, which are a function of the degree of polymerization. Following the theory that the existence of the LC state is explained only by the packing effects of the anisometric molecules, the LC phase stability is determined by the axial ratio of the rigid unit[3]. Consequently for a rigid-rod polymer such as that shown in Figure 6.2(a), the LC phase stability should be directly determined by the degree of polymerization (X), whereas for the other model, shown in Figure 6.2(b), only the mesogenic element in the backbone should determine the phase stability.

In reality, polymers exhibit behaviour which resembles that of both models. The degree of flexibility of the main chain determines the phase stability, and also, as for LLC, not only the axial ratio of the mesogenic groups, but also their chemical constitution and their substituents strongly affect the LC phase behaviour. This will be discussed later.

6.2.1 *Synthesis*

The synthesis of LC main chain polymers normally follows step-growth reactions, such as polycondensation or polyaddition, where the reactants contain two monofunctional groups. The monomers can either be a suitably substituted mesogenic compound or a non-mesogenic substance. These monomers form the mesogenic segment of the polymer main chain by a polymerization reaction. To obtain high molar mass polymers, the functional groups are limited to those giving reactions of high yield[8]. Polyesters or polyamides are preferred systems, because high molar mass polymers can be easily obtained by standard polymerization processes.

(a) Rigid main chain. A large number of polymers having a rigid, rod-like main chain have been prepared[9, 10] and many are described in the patent literature. Commercially easily available materials like *p*-hydroxybenzoic acid (**1**) are preferred as starting materials. Although low molar mass phenyl esters of benzoic acid form a large class of intensively investigated LC materials, the polycondensation of compound **1**:

$$x\ \mathrm{HO{-}C_6H_4{-}COOH}\ (\mathbf{1}) \xrightarrow[-(x-1)\mathrm{H_2O}]{} \mathrm{H}{-}\left[\mathrm{O{-}C_6H_4{-}\overset{O}{\overset{\|}{C}}}\right]_x{-}\mathrm{OH}$$

and the polycondensation of terephthalic acid and hydroquinone:

$$x\ \mathrm{HO{-}C_6H_4{-}OH} + x\ \mathrm{HOOC{-}C_6H_4{-}COOH} \xrightarrow[-(2x-1)\mathrm{H_2O}]{}$$

$$\mathrm{H}{-}\left[\mathrm{O{-}C_6H_4{-}O\overset{O}{\overset{\|}{C}}{-}C_6H_4{-}\overset{O}{\overset{\|}{C}}}\right]_x{-}\mathrm{OH}$$

yield polyesters which are not LC materials[11]. The polymers are insoluble and do not fuse without the occurrence of decomposition. They can be processed only by sintering.

The polyamide of terephthalic acid and 1,4-phenylenediamine:

$$\mathrm{HOOC{-}C_6H_4{-}COOH} + \mathrm{H_2N{-}C_6H_4{-}NH_2} \longrightarrow \mathrm{HO{-}\!\left[\overset{O}{\overset{\|}{C}}{-}C_6H_4{-}\overset{O}{\overset{\|}{C}}{-}NH{-}C_6H_4{-}NH\right]\!{-}H}$$

which is well known under the trade name 'Kevlar', is also infusible, but is soluble in concentrated sulphuric acid. In solution, a nematic phase occurs which is processible, for example, by fibre spinning.

These examples demonstrate that rigid, rod-like polymers having a symmetrical chemical constitution are not suitable for obtaining LC polymer melts. With increasing X, beginning with the monomer, the melting temperature rapidly increases above the decomposition temperature. To overcome the problems posed by the high temperatures required for the transformation from the crystalline to the LC phase, two modifications of the polymer main chain are conceivable. Firstly, the cylindrical shape of the backbone can be reduced by introducing non-linear but rigid elements in the main chain. With increasing amounts of the non-linear comonomer unit, the melting point of the copolymer falls below the decomposition temperature and LC phases may be observed. Some typical examples of copolymers containing non-linear monomer units are shown in Table 6.1. The second possibility is given by substituting lateral alkyl chains, which act as lubricants between the rod-like particles, onto the mesogenic mononer units. The lateral, flexible substituents can also be considered as 'solvent molecules' attached to the macromolecule. This concept has been successfully applied to cellulose derivatives, which become thermotropic LC materials with long-chain, lateral substituents[18].

(b) Semiflexible main chain. To overcome the problems of the high melting points of polymers containing only rod-like chain segments, flexible moieties like alkylene or alkyleneoxy chains can be introduced into the polymer main chain. Again, a variety of systems is conceivable by varying the chemical structures of the mesogenic moieties, the lengths of the flexible segments (flexible spacers) or the number of different monomer units in regular or statistically irregular copolymers. Although an impressive volume of work has been done on these systems and has been reviewed[10, 19], we will restrict ourselves to three typical examples.

Polymerization of α,ω-substituted polymethylene chains with functionalized mesogenic compounds provides a convenient method for making semiflexible main chain polymers. Polyesters synthesized from 4,4′-dihydroxy-2,2′-dimethylazobenzene and alkandioic acids:

$$x\ \mathrm{HO{-}C_6H_3(CH_3){-}N{=}N{-}C_6H_3(CH_3){-}OH} + x\ \mathrm{HOOC{-}(CH_2)_n{-}COOH} \longrightarrow \left[\mathrm{O{-}C_6H_3(CH_3){-}N{=}N{-}C_6H_3(CH_3){-}OOC{-}(CH_2)_n{-}CO}\right]_x$$

exemplify such a system[21].

Another approach is the polymerization of a non-mesogenic monomer such that a rigid, but flexibly linked, mesogenic moiety is built up by a polycondensation reaction:

$$x\ \mathrm{HO{-}C_6H_3(R){-}OH} + x\ \mathrm{HOOC{-}C_6H_4{-}O{-}(CH_2)_n{-}O{-}C_6H_4{-}COOH} \longrightarrow \left[\mathrm{(CH_2)_n{-}O{-}C_6H_4{-}COO{-}C_6H_3(R){-}OOC{-}C_6H_4{-}O}\right]_x$$

or

$$x\ \mathrm{HO{-}C_6H_4{-}O{-}(CH_2)_n{-}O{-}C_6H_4{-}COOH} \longrightarrow \left[\mathrm{(CH_2)_n{-}O{-}C_6H_4{-}COO{-}C_6H_4{-}O}\right]$$

An interesting synthesis has been described by Jackson and Kuhfuss[22]. The starting material is the commercially well-known poly(ethyleneterephthalate) (PET), which is transesterified with *p*-hydroxybenzoic acid (PHB). With increasing amounts of PHB, the mesogenic moieties are statistically irregularly built up in the main chain. Furthermore, the mesogenic groups differ in their length:

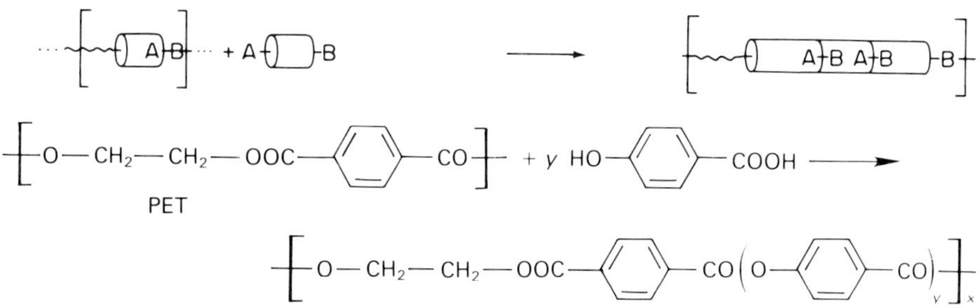

$$\cdots\sim[\,A\text{—}B\,]\cdots + A\text{—}B \longrightarrow [\sim A\text{—}B\ A\text{—}B\text{—}B\,]$$

$$\left[\text{O—CH}_2\text{—CH}_2\text{—OOC—C}_6\text{H}_4\text{—CO}\right] + y\ \text{HO—C}_6\text{H}_4\text{—COOH} \longrightarrow$$

PET

$$\left[\text{O—CH}_2\text{—CH}_2\text{—OOC—C}_6\text{H}_4\text{—CO}\left(\text{O—C}_6\text{H}_4\text{—CO}\right)_y\right]_x$$

Table 6.1 Selected examples of rigid, non-linear LC copolymers

	References
$\left(\text{O—C}_6\text{H}_2(X)(Y)\text{—O—C(=O)—C}_6\text{H}_4\text{—C(=O)}\right) \sim \left(\text{O—C}_6\text{H}_2(X)(Y)\text{—O—C(=O)—Z—C(=O)}\right)$ X, Y = —H, —Cl, —Br, —CH_3 Z = —C_6H_{10}(H)—, —C_6H_4—O—C_6H_4—, 2,6-naphthylene	12, 13, 14
—O—$C_6H_3(C_6H_5)$—O—C(=O)—C_6H_4—C(=O)—	15
$\left(\text{C(=O)—C}_6\text{H}_4\text{—O—C(=O)—C}_6\text{H}_4\text{—C(=O)—O—C}_{10}\text{H}_6\text{—O}\right)$ and $\left[\text{O—C}_6\text{H}_4\text{—O}\right]$ or $\left[\text{O—C}_6\text{H}_4\text{—C(=O)}\right]$	16
—O—C_6H_4—C(=O)—O—$C_{10}H_6$—C(=O)—	17

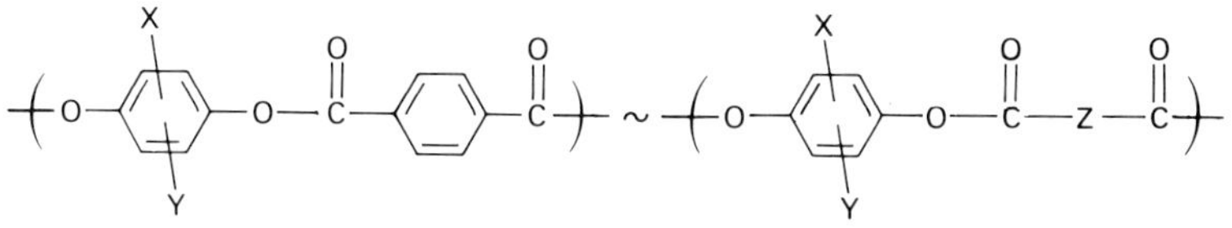

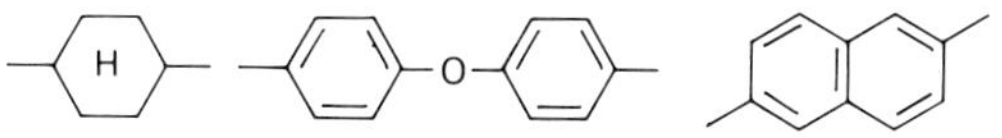

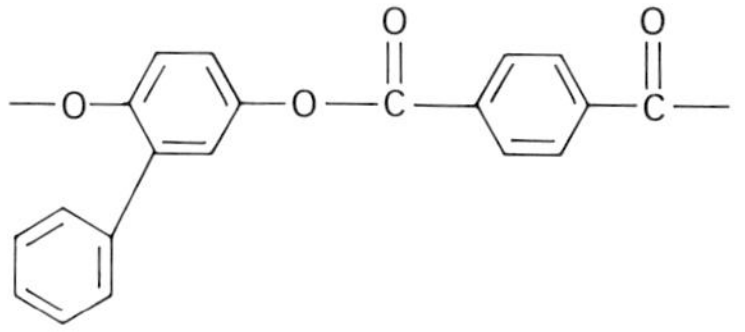

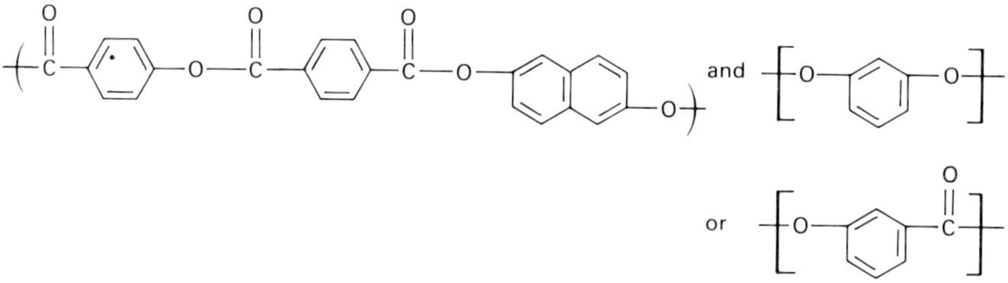

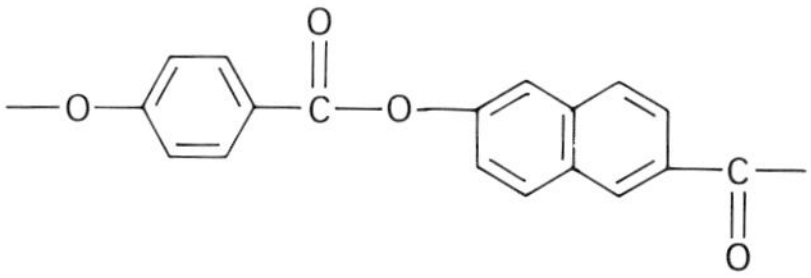

By this method, the flexibility of the main chain can be regulated by the amount of transesterified PHB.

6.2.2 *Phase behaviour in relation to molecular structure*

For LC main chain polymers, the existence of nematic (N), cholesteric (Ch) and smectic (S) phases has been found. In contrast to LLC materials where, in a homogeneous LC phase of one component, all molecules have the same size or molar mass, the relation between phase behaviour and molecular structure is more complex for LC main chain polymers. The polymers have a molar mass distribution which is determined by the polymerization process. If we consider rigid-rod polymers, the occurrence of an N phase can be easily understood (Figure 6.3a).

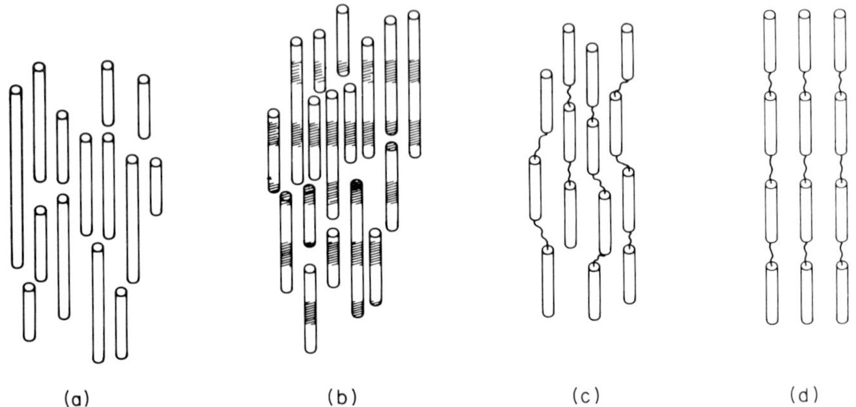

Figure 6.3(*a*) and (*b*) Nematic and smectic phases of rigid-rod polymers. (*c*) and (*d*) Nematic and smectic phases of semiflexible polymers.

Because of the length and molar mass distribution of the rod-like macromolecules, their centres of gravity become statistically arranged, causing the N order. An S phase is only conceivable for polymers which have a regular sequence of monomer units, such that the centres of the monomer units can become arranged in layers (Figure 6.3b). Hence, it follows directly that statistically irregular copolymers, non-linear backbones, or laterally substituted mesogenic monomer units suppress the formation of S phases and promote the occurrence of N phases.

If we consider semiflexible polymers, a phase sequence more similar to LLC materials is expected. The flexible spacer between the mesogenic moieties can be regarded as very similar to the terminal alkyl substituents commonly found in LLC. In the case of homopolymers having a regular sequence of rigid, mesogenic moieties and flexible spacers, the structure of the LC phase is determined by the mesogenic moieties within the main chain (Figure 6.3c and d). Short, flexible spacers favour N phases, while longer spacers favour smectic-like packing. Statistically irregular copolymers, where the lengths of the mesogenic moieties or the lengths of the flexible spacers vary, prefer the N ordering.

As already mentioned, LC main chain polymers are normally crystalline or at least partially crystalline up to high temperatures, before the appearance of the LC state. Because of the polydispersity of the polymers, broad crystalline-to-LC phase transformation regimes, where crystalline and LC polymer coexist, are observed. The same holds for the LC-to-isotropic phase transformation. Biphasic regions can be found over a temperature interval of up to some ten degrees. For nearly all polymers described in the literature, definite temperatures are given although the phase transformations are broadly smeared out because of the polydispersity of the polymers. In most cases the transformation temperature is simply given as the temperature of maximum heat flow (determined from differential scanning calorimetric curves) without presenting any information about the width of the biphasic region. Detailed information about the width of the biphasic region has been presented only by Blumstein[23] for the semiflexible polyester shown in Figure 6.4. For polymers having a heterogeneity index, $\bar{M}_w/\bar{M}_n$, of about 2.1–2.2, the biphasic region strongly depends on the molar mass and becomes more narrow with increasing degree of polymerization, X_n. This relation, however, holds only for semiflexible polymers and has not been investigated for rigid-rod polymers.

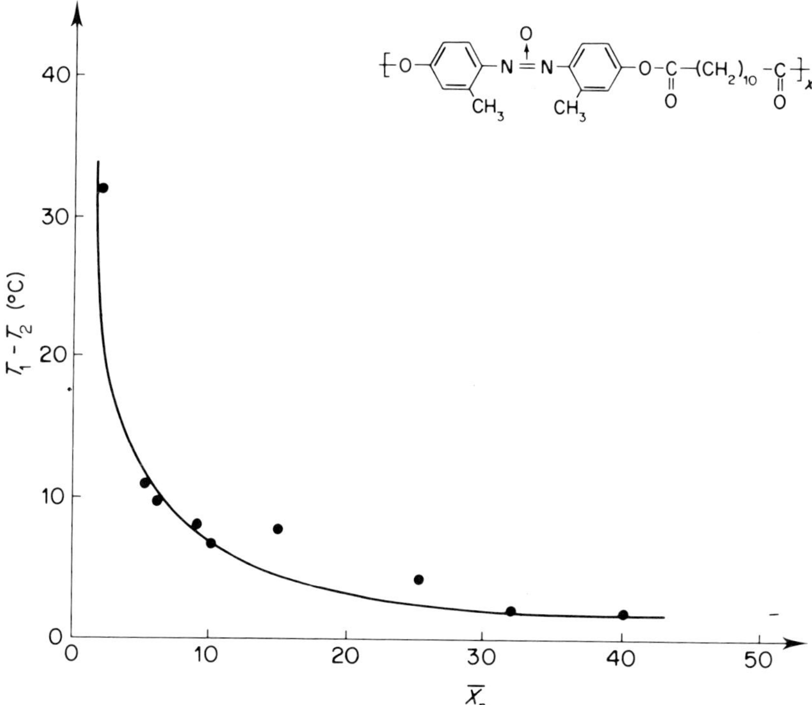

Figure 6.4 Biphasic gap (T_1–T_2) for the LC-to-isotropic phase transformation as a function of the degree of polymerization X.

(a) Rigid-rod polymers in bulk. Investigations into the relationship between phase behaviour and molecular structure have concentrated mainly on the crystalline-to-LC phase transformation rather than on the LC-to-isotropic liquid transformation, because the latter is expected to occur above the decomposition temperature of the polymer. For LLC, well-defined relations can easily be obtained between chemical constitution and phase behaviour because of the well-defined purity and molar mass of the materials. The situation becomes complex for polymers. Strictly speaking, only polymers of the same average degree of polymerization and the same molar mass distribution can be directly compared. Furthermore, the crystalline-to-LC transformation depends on the degree of crystallinity of the polymer and the size distribution of the crystalline particles. Phase transformation enthalpies and temperatures are therefore path dependent and strongly depend upon thermal history. Consequently, comparisons of thermodynamic data for different polymers can reveal no exact relation between phase transformation temperatures and chemical constitution because information relating to the above relevant factors is simply not available in the literature for rigid-rod polymers. A further severe problem in analysis is the poor solubility of the polymers. Only exotic solvents, such as concentrated sulphuric acid for polyamides or substituted phenols for polyesters, are known. With the exception of methods involving viscosity measurements, conventional methods for the determination of molar mass have been unsuccessful to date. The method involving viscosity measurements is, however, a relative method, yielding no absolute values without calibration. Therefore only approximate relations between phase behaviour and molecular structure can be made. These are mentioned in Section 6.2.1.

(b) Rigid-rod polymers in solution. The addition of a suitable isotropic, low molar mass solvent to a rigid-rod polymer depresses the crystalline-to-LC as well as the LC-to-isotropic phase transformation temperature. By this method, LC phases can be observed for polymer solutions, for which the anisotropic phase of the pure system lies above the decomposition temperature. For LLCs, with increasing concentration of the solvent, the LC-to-isotropic phase transformation temperature decreases, and at a certain concentration, the LC phase no longer forms. Only the crystalline state and/or the isotropic solution exist.

On the other hand, when beginning with pure solvent and adding increasing amounts of polymer, an LC phase appears at a certain concentration and separates out from the isotropic solution. Further addition of the polymer yields a homogeneous, anisotropic solution. The lowest concentration at which nematic droplets separate from the isotropic solution is called the critical concentration and is theoretically predicted, based on a variety of approaches. Flory's calculation[3, 24], which is based only on packing effects of monodisperse, rigid rods, yields

$$V_{\text{crit}} = \frac{8}{x}\left(1 - \frac{2}{x}\right)$$

where V_{crit} is the critical volume concentration and x is the axial ratio of the rods, which is directly proportional to the degree of polymerization X or the molar mass of the polymer. Increasing X decreases the critical concentration. This is qualitatively observed for many polymer-solvent systems. On the other hand, increasing the flexibility of the polymer backbone increases the polymer concentration required for observation of the anisotropic phase. With increasing polymer concentration at constant temperature, the following sequence of phases appears: isotropic solution; biphasic regime of isotropic solution and anisotropic solution; homogeneous anisotropic solution. The concentration range of these regimes depends on the molar mass, the stiffness and the polydispersity of the macromolecule. Furthermore, solution behaviour is dependent on the solvent.

For these (quasi-)binary, solvent-polymer mixtures, some confusing nomenclature exists in the literature. Because the LC phase is observed only for a solution, owing to the low decomposition temperatures of polymers, these mixtures are often called 'lyotropic polymers'. However, it is well established that lyotropic, LC phases are formed by solutions of amphiphilic molecules, whose micellar aggregates in solution become anisotropically ordered in defined concentration and temperature ranges — a rather different situation.

(c) Semiflexible polymers. The characterization of semiflexible polymers, which contain flexible spacers between the rigid chain segments, is less difficult, because these macromolecules are normally soluble in common solvents and their LC-to-isotropic phase transformation temperatures are below their decomposition temperatures. The polymers can be separated with respect to molar mass, and their physicochemical properties are available as functions of X. This has been intensively investigated by Blumstein[21, 23], who has observed that, beginning with the monomer, as X increases, the LC-to-isotropic phase transformation temperature increases and remains unchanged for $X > 10$ (Figure 6.5). The same relation holds for the phase transformation entropy, $\Delta S_{N,I}$. These results clearly indicate that the flexible spacer in the polymer backbone strongly enhances the mobility of the rigid, mesogenic chain segments. Furthermore, from these results, assuming a sufficient spacer length, the polymers do not change their LC phase behaviour when $X > 10$, whereas with decreasing spacer length, the phase transformation temperature has to be related to X for polymers with $X < 10$.

Besides the degree of polymerization, the phase behaviour can be systematically influenced by the length of the flexible spacer and the chemical constitution of the mesogenic moiety[23, 25, 26].

(i) Spacer length. As already mentioned, increasing the length of the flexible spacer increases the mobility of the mesogenic groups. Decreasing LC-to-isotropic phase transformation temperatures and an increasing tendency towards the formation of smectic phases should result. These effects are confirmed[26] by the phase behaviour of a polyester, as a function of the spacer length, as shown in Figure 6.6. A strong 'odd–even' effect, which is well known for LLCs, is observed.

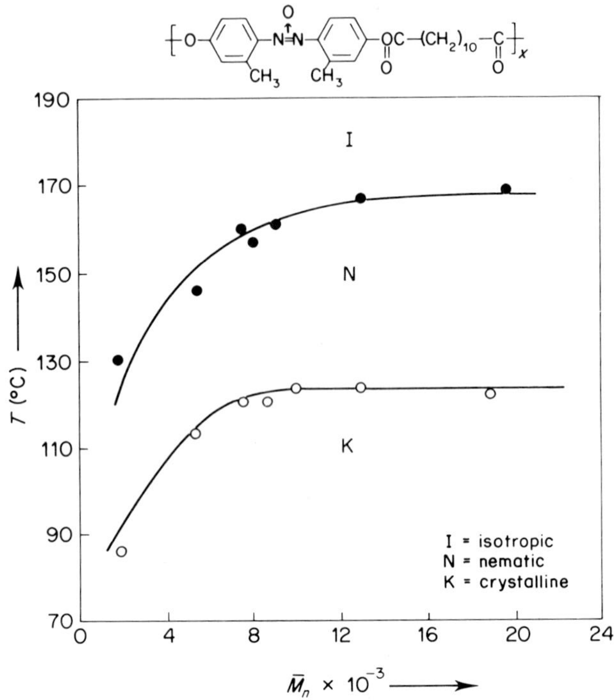

Figure 6.5 Temperature dependence of the phase transitions as a function of the molar mass M_n of the semiflexible polymers.

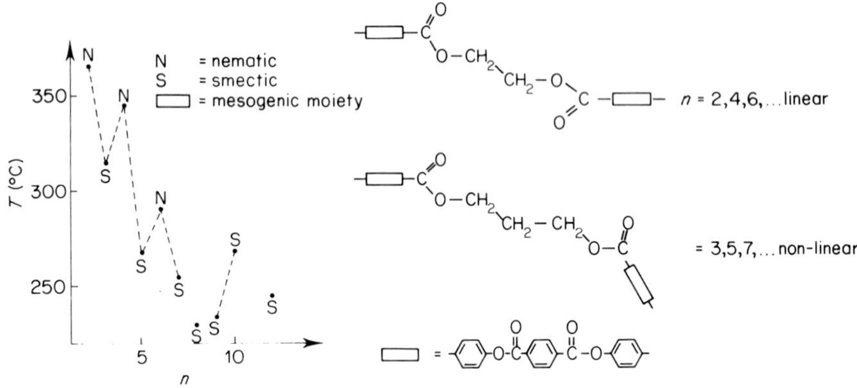

Figure 6.6 Phase transformation temperatures for polyesters with different spacer lengths, n.

An even number of atoms in the spacer causes high transformation temperatures, and an odd number, low transformation temperatures. This behaviour can be explained if we consider a fully extended *trans* conformation of the spacer. A linear or non-linear arrangement of the mesogenic groups occurs, and the non-linear arrangement for the odd number of atoms reduces the stability of the mesophase. The odd–even effect is also reflected in the phase transformation entropies and enthalpies.

It has to be mentioned that a chiral carbon atom can be easily introduced into a flexible spacer. Presuming the non-chiral polymer is N, the chiral centre converts the LC phase of the polymer into a Ch phase[27]. For Ch polymers, an increasing concentration of the chiral component decreases the helicoidal pitch; this relationship is known for LLCs and for mixtures of nematics and chiral components.

(ii) Mesogenic moieties. Small alterations of the mesogenic moiety can strongly affect the LC phase behaviour of the polymers, thus making prediction of properties nearly impossible. Nevertheless, some tendencies valid for LLCs have been established. Increasing length and stiffness of the mesogenic moieties within the backbone increases the thermal stability of the LC phase. Deviation from a cylindrical shape brought about by non-linear moieties or lateral substituents reduces the LC-to-isotropic phase transformation temperature. Interestingly, polar substituents, e.g. the exchange of a —CH_3 group for a polar —Cl or —Br within an homologous series has little influence on LC phase behaviour[28]. This indicates that mainly packing effects determine phase stability. Long, pendent, flexible, lateral groups like n-alkyl chains also strongly reduce the LC phase stability; this is demonstrated for the polymer series[26]:

$$\left[-O\overset{O}{\overset{\|}{C}}-C_6H_4-\overset{O}{\overset{\|}{C}}-O-C_6H_3\big((CH_2)_n-CH_3\big)-O-\overset{O}{\overset{\|}{C}}-C_6H_4-\overset{O}{\overset{\|}{C}}-O-(CH_2)_{10}- \right]$$

The behaviour is similar to that found for the corresponding low molar mass phenyl esters of hydroquinone[29].

6.2.3 *Properties and aspects relating to applications*

The most important property of LC main chain polymers in the N state, in bulk or in solution, is the ease of homogeneous orientation of the molecular long axes upon the application of shear. This macroscopic orientation gives rise to a viscosity lower than that of the isotropic state, despite the lower temperature of measurement in the former case. The steady shear viscosity is shear-rate dependent and is higher at low frequency or low shear rate because of incomplete orientation of the polymer[30]. Therefore extrusion or fibre spinning of these polymers in the N state is of technological importance. Processing temperatures can be reduced, and mold filling would be improved.

In solidified extrudates or fibres, the material may be glassy, partially crystalline or highly crystalline. The macroscopic alignment of the macromole-

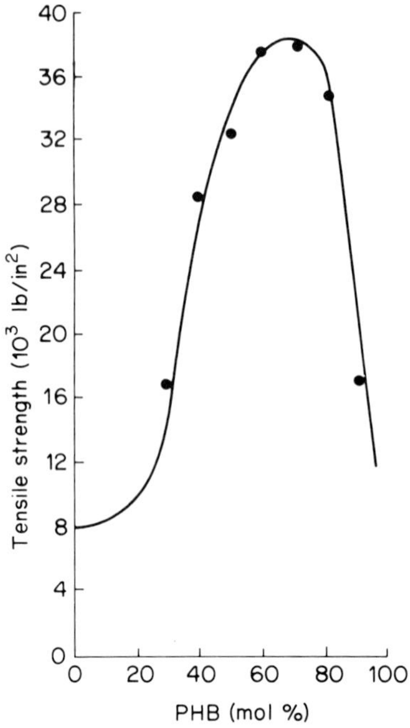

Figure 6.7 Tensile strength of poly(ethyleneterephthalate) transesterified with *p*-hydroxybenzoic acid.

cules due to the orientation during processing remains and creates anisotropic physical properties.

The most important factors with respect to applications are the outstanding mechanical properties, e.g. tenacities are much higher than those of conventional spun yarns. An interesting example is shown in Figure 6.7, where the tensile strength for the polymer from poly(ethyleneterephthalate) transesterified with *p*-hydroxybenzoic acid (PHB) is shown as a function of the concentration of PHB in the copolyester[22]. Although the tensile strength remains nearly unchanged up to about 30% PHB in the macromolecule, it then strongly increases, beginning with the concentration at which the polymers become liquid crystalline.

The macroscopic orientation of the macromolecules improves mechanical properties like tensile strength, flexural strengh, flexural moduli and impact strength only along the direction of orientation. Across this direction, the mechanical properties are poorer. This is illustrated in Figure 6.8, which shows the typical wood-like structure of an LC main chain polymer and its anisotropic properties at the break. The absence of shrinking, low coefficients of thermal expansion and outstanding mechanical properties are attractive properties of LC main chain polymers for their use as fibres, extrudates and reinforced resins.

Besides the mechanical properties, two other important properties have to be

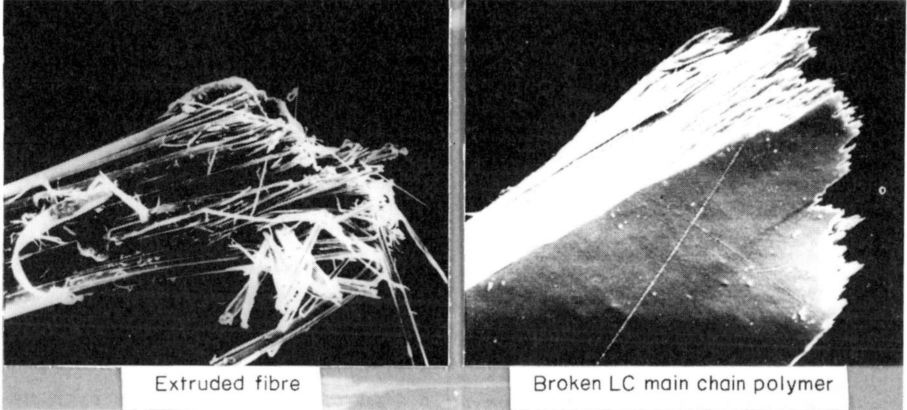

Figure 6.8 SEM (scanning electron microscopy) photograph for an extruded fibre and film of an LC main chain polymer after breaking. (Photo kindly placed at the author's disposal from BASF, Ludwigshafen, West Germany.)

mentioned. Although LC main chain polymers can be easily macroscopically ordered in the N phase by flow, electric or magnetic field-induced orientation fails for polymers having high molar mass[31, 32]. However, with sufficiently low degrees of polymerization (X), field effects can be observed by analogy with LLCs. Systematic investigations, e.g. with respect to elastic constants, have not yet been made. Another important physical property is the state of order in N main chain polymers; this is also of theoretical interest[33]. Experimentally it has been confirmed that with increasing X, the N order parameter increases above that obtained for the chemically related LLCs[23].

6.3 Liquid crystal side chain polymers

The anisotropic LC state first attracted the attention of polymer chemists two decades ago. There were two reasons for this: the possibility of using the LC state as a model for studying the kinetics of polymerization of anisotropically ordered molecules and the influence of the ordered state on the tacticity of the growing chain. Many difficulties arose in interpreting the results because of the inhomogeneity of the reaction mixture. For example, although the monomers exhibited an N state, the growing polymers became isotropic or smectic, but not N. Phase separation occurred, and the polymerization process was controlled by diffusion, not by the anisotropic order. One principal problem became obvious: if the polymerizable group is directly attached to the rigid mesogenic monomer, the resulting polymer does not necessarily exhibit the LC state. In most cases an amorphous polymer is obtained. These problems can be overcome by a simple concept. If the rigid mesogenic molecules are attached to the polymer backbone via a 'flexible spacer', e.g. an alkylene or alkyleneoxy chain, LC phases can be realized[34, 35]. This is based on a relationship between chemical constitution and LC behaviour similar to that for LLCs, as described in Chapter 2, and can be explained by (at least) partial decoupling of the motions between the mesogenic side chain and the chain segments of the backbone. Although above the glass transition temperature the polymer main chain tends to adopt a statistical chain conformation, the flexible spacer enables the mesogenic side chains to order anisotropically. To a first approximation, the statistical coil conformation is consistent with an N phase, where the mesogenic side chains are statistically disordered with respect to their centres of gravity. Measurements indicate, however, that the macromolecular chain deviates from the statistical coil conformation and becomes anisotropically deformed[36]. For S phases, the chain conformation has to adapt to the layered arrangement of the mesogenic side chains. A two-dimensional ordering between the smectic layers is a possible explanation[37].

6.3.1 Synthesis

The synthesis of LC side chain polymers follows well-known polymerization procedures established in polymer chemistry. Step-growth reactions (e.g. polycondensation or polyaddition reactions) or chain-growth reactions (e.g. radical or ionic polymerization) can be applied. Limitations exist only with respect to side reactions, such as chain transfer reactions or reactions of the initiator with functional groups of the mesogenic unit of the monomer. Therefore chain growth reactions are mainly limited to radical-initiated polymerization. Some typical examples will now be described for the synthesis of LC side chain polymers.

(a) Chain-growth reactions. The most frequently applied polymerization reaction is the radical-initiated polymerization of mesogenic monomers. Because of their ease of polymerization, preferred systems are derivatives of acrylic acid or methacrylic acid:

H(CH$_3$)
O—(CH$_2$)$_n$—[ ] ⟶ [CH$_2$—C(H(CH$_3$))(C=O—O—(CH$_2$)$_n$—[ ])]
O

[ ] = mesogenic moiety

n = 0,1,2,3, ... length of flexible spacer

Typical mesogenic moieties are derived from substituted benzoic acid phenyl esters, aromatic azo or azoxy compounds, aromatic Schiff bases, substituted biphenyls and cholesterol. They are comprehensively summarized in the literature[35, 38, 39]. In most cases, polymerization is performed in solution and initiated with common initiators such as azobisisobutyronitrile. Although high molar mass polymers can be easily obtained (mol. wt $\approx 10^6$ g/mol) for methacrylic acid derivatives, chain transfer reactions of acrylic acid derivatives normally yield polymers with lower molar mass.

Ionic polymerization of LC monomers is limited to systems that survive the reaction steps of initiation and chain propagation without appreciable side reactions. Anionic polymerization of a biphenyl-based methacrylate has been mentioned, but yields only polymers with low molar mass[40, 41]. The ionic polymerization of ethylene oxide-containing LC monomers also yields only oligomers[42].

Besides homopolymerization, copolymerization of two or more monomeric components affords a broad variation in the preparation of LC polymers. The copolymerization of mesogenic, non-mesogenic and chiral components enables the phase behaviour or phase structure of the copolymers to be varied. This change in phase behaviour is similar to that which occurs when mixing LLC materials. Assuming identical reactivity of each of the monomeric components in the polymerization reaction, the monomer composition of the copolymer is identical to the composition of the monomer mixture.

(b) Step-growth reactions. Although step-growth reactions are the basis for the synthesis of LC main chain polymers, this polymerization process has not been applied to LC side chain polymers, except in the polycondensation of a substituted malonic acid with a diol[43]:

$$\mathrm{HOOC{-}\underset{\underset{\underset{\Box}{|}}{\underset{O}{|}}}{\underset{(CH_2)_n}{\underset{|}{CH}}}{-}COOH + HO{-}Z{-}OH \longrightarrow \left[Z{-}OOC{-}\underset{\underset{\underset{\Box}{|}}{\underset{O}{|}}}{\underset{(CH_2)_n}{\underset{|}{CH}}}{-}COO \right]}$$

$$\Box = -\mathrm{C_6H_4{-}N{=}N{-}C_6H_4{-}CN}$$

$$\mathrm{Z} = -(\mathrm{CH_2})_9-,\ -\mathrm{CH_2{-}C_6H_4{-}CH_2}-$$

The polyesters exhibit LC properties and offer a convenient method for the preparation of polymer main chains containing heteroatoms.

(c) Modification of polymers. Reactions with polymers containing functional groups in the monomer unit can also be applied to the synthesis of LC side chain polymers. Problems occur when the reactions are not quantitative or are accompanied by side reactions. Steric hindrance arising from bulky mesogenic groups may also cause difficulties. Well-defined LC polymers can be obtained by

this type of procedure using the reaction of a poly(hydrogenmethylsiloxane) with a mesogen containing a vinyl group[44–46]:

$$CH_2{=}CH{-}(CH_2)_{n-2}{-}\square \; + \; {-}\!\left[\overset{CH_3}{\underset{H}{Si}}{-}O\right]\!{-} \longrightarrow {-}\!\left[\overset{CH_3}{\underset{(CH_2)_n-\square}{Si}}{-}O\right]\!{-}$$

The addition reaction is catalysed by platinum catalysts, the amount of which must be optimized for this reaction. Pentadienyl complexes with platinum give high yields and only minor side reactions. The LC poly(siloxanes) are of special interest because of their low glass transition temperatures.

The reaction of poly(acryloyl chloride) with mesogenic molecules containing a suitable —OH group[47], as well as the addition reaction between poly(sodium acrylate) and mesogenic ω-bromoalkyl esters under phase transfer conditions[48] have successfully been applied for the synthesis of LC side chain polymers.

6.3.2 Phase behaviour in relation to molecular structure

LC side chain polymers are characterized by a combination of polymer-specific and LC-specific properties. The polymers may exhibit N, Ch (chiral N) and S phases similar to LLCs.

Comparing the phase behaviour of the mesogenic monomer with that of the corresponding polymer, two important changes in property are observed: (1) the increased tendency towards formation of S phases and (2) the increased LC phase transformation temperatures for the polymers[49].

The tendency towards formation of S phases for the polymers might be explained by the reduced mobility of the mesogenic side chains, which are connected via the polymer backbone. Rotational as well as translational motions are restricted, favouring the formation of the S state. Particularly in the absence of flexible spacers, where the bulky mesogenic moieties are directly attached to the main chain, either no LC phase is obtained because of steric hindrance or an S polymer results. These S polymers are characterized by high clearing temperatures, T_c, and high glass transition temperatures, T_g[50]. The high T_g values of these polymers compared with those of the corresponding unsubstituted polymers arise due to the strong stiffening of the backbone by the mesogenic moieties. These results indicate that for the production of N polymers, a sufficient mobility of the mesogenic side chains is necessary through having either a flexible spacer or a sufficient flexibility of the chain segments.

The increase of the LC phase transformation temperatures by polymerization can be almost quantitatively explained by the reduction of the specific volume in going from the monomer to the polymer[49]. A strong decrease in specific volume is observed for the oligomers ($X < 10$) and levels off for $X > 100$. Similarly,

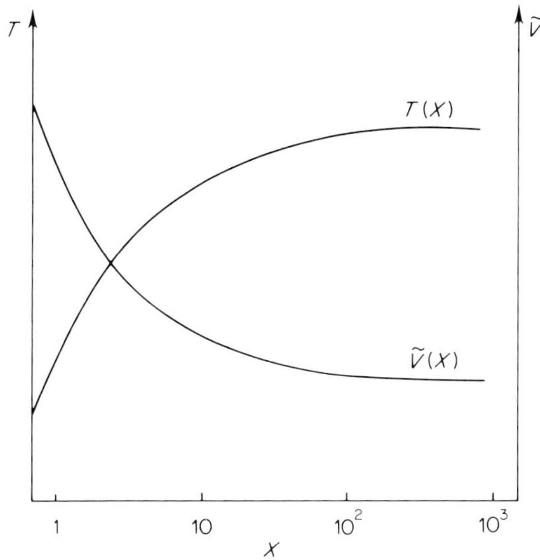

Figure 6.9 Schematic presentation of the phase transformation temperature T and the specific volume $\tilde{V}$ as function of the degree of polymerization X for LC side chain polymers.

the phase transformation temperatures increase with X, as shown schematically in Figure 6.9. Interestingly, the S–N or S–I transformation temperatures do not increase with X as strongly as the N–I transformation temperatures:

$$\frac{\Delta T_{\mathrm{S-N}}}{\Delta X} \sim \frac{\Delta T_{\mathrm{S-I}}}{\Delta X} < \frac{\Delta T_{\mathrm{N-I}}}{\Delta X}$$

The same relations hold for LLCs, where the specific volume can be reduced by pressure.

A third important aspect of LC side chain polymers is their glass transition at low temperatures; this is mainly determined by the chemical constitution of the polymer main chain. Although LLCs will normally crystallize at or just below T_{m} on lowering the temperature, polymers can be easily supercooled. This is shown schematically in Figure 6.10 where the specific volume is given as a function of the temperature. At the glass transition temperature, T_{g}, polymers vitrify, to yield glasses which have an LC structure. The glass transition can be identified by a bend in the $T(V)$ curve. In Figure 6.10, a glassy polymer with an S structure is obtained at $T_{\mathrm{g,S}}$. The N phase can also be vitrified at $T_{\mathrm{g,N}}$ by rapid cooling. The formation of anisotropic glasses is one of the most important and interesting properties of LC side chain polymers. This leads to broad aspects of applications mentioned in Section 6.3.3. When annealing the supercooled LC melt at temperatures T, where $T_{\mathrm{g,N}}, T_{\mathrm{g,S}} < T < T_{\mathrm{m}}$, LC polymers in many cases become partially crystalline[51]. This is indicated by the jump in the $V(T)$ curve.

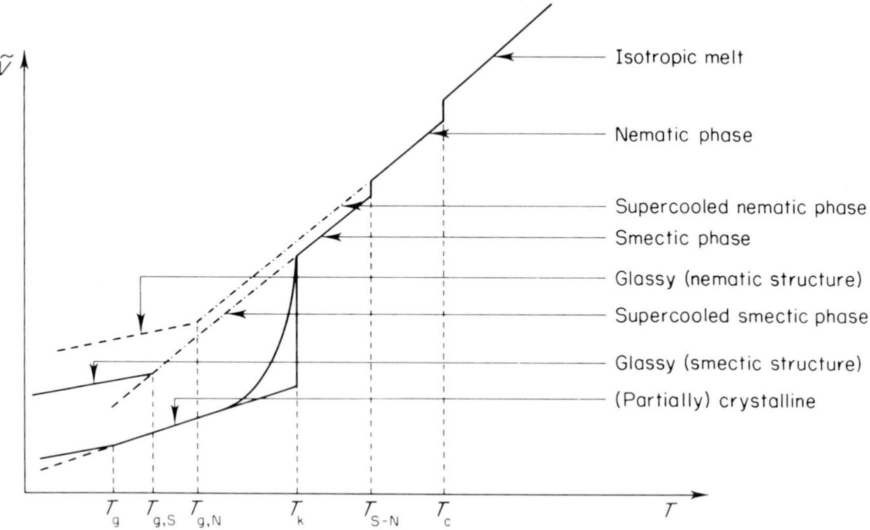

Figure 6.10 Schematic representation of the specific volume $\widetilde{V}$ versus temperature for LLC materials and LC polymers.

In addition to the properties mentioned above, investigations indicate a relationship between phase behaviour and molecular structure very similar to that mentioned in Chapter 2 for LLCs. Some typical examples will be mentioned.

(a) Substituents at the mesogenic moiety. Within a homologous series, increasing the length of the linearly arranged units of the mesogenic moiety favours the appearance of S phases[34, 46]. This is also observed for LC side chain polymers (Table 6.2b). We must consider not only the length of the mesogenic moiety and of the flexible spacer but also that of the chain segment to the next mesogenic monomer unit. This means that an N homopolymer can be converted into an S copolymer by inserting monomer units without mesogenic groups into the polymer main chain[52, 53]. Lateral substituents on the mesogenic moiety strongly depress the clearing temperature, especially for polymers with short flexible spacers[46]. This indicates strong restrictions in packing possibilities due to the linkage to the backbone. The introduction of chiral substituents or the copolymerization of a nematogenic monomer with a chiral comonomer yields cholesteric phases[34, 54, 55].

Although the systematic relationships obtained for LLCs (see Chapter 2) can be applied to LC side chain polymers which have *no strongly polar substituents*, strong deviations arise with polar mesogenic moieties. With a few exceptions[46, 56], these polymers are *smectic*. Up to now, no experimental results have provided a detailed explanation. It can be assumed, however, that because of the polar substituents, pair formation between mesogenic groups

Table 6.2 Phase transition temperatures for selected (a) polar[46] and (b) non-polar LC side chain polymers[34]

(a) $-[Si(CH_3)-O]-$ with side chain $(CH_2)_m-O-C_6H_4-COO-C_6H_4-CN$

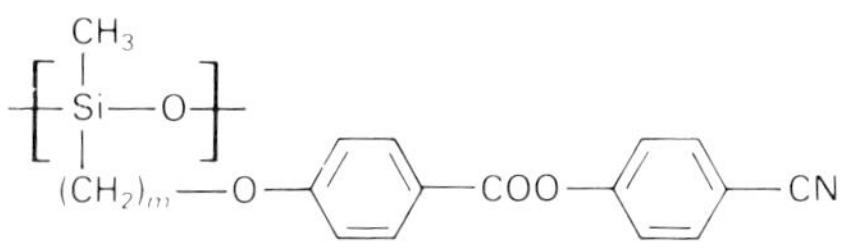

(b) $-[Si(CH_3)-O]-$ with side chain $(CH_2)_m-O-C_6H_4-COO-C_6H_4-OCH_3$

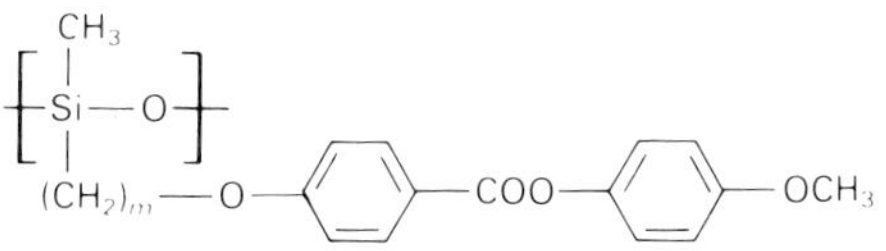

m	Phase transformation temperatures [°C] (a)	(b)
3	g 29 S 161.5 I	g 15 N 61 I
4	g 26 S 150 I	g 15 N 102 I
5	K 107 S 183 I	(g 17) K 87 N 115 I
6	K 55 S 181 I	(g 5 S 46) K ~ 57 N 108 I
7	K 88 S 189 I	
8	K 62 S 194 I	
11	K 76 S 198 I	

occurs, thereby stiffening the backbone and reducing the occurrence of N ordering. This argument is supported by LC side chain polymers in which the mesogenic group is laterally attached to the polymer backbone (see Figure 6.11). For these systems, N phases occur with polar mesogens[57]. Here, pair formation does not restrict main chain mobility, because the mesogenic groups are less restricted in their motion about the lateral linkage to the backbone.

Another interesting result is that LLCs exhibit a distinct odd–even alternation of their clearing temperatures (T_c) with respect to the length of the terminal chain. This odd–even effect has also been observed for S side chain polymers, but not for analogous series of N polymers (Table 6.2). If the mesogenic groups are laterally attached to the backbone, a pronounced odd–even effect is observed for N polymers (Figure 6.11), by analogy with LLCs. This might indicate a smaller influence of the backbone on the anisotropic arrangement of the mesogenic units, as mentioned above.

(b) Chemical constitution of the polymer main chain. The chemical constitution of the polymer main chain determines the glass transition temperature, T_g. Increasing the flexibility of the backbone depresses T_g. This also holds for LC side chain polymers, assuming the same mesogenic group is attached via the same flexible spacer to the main chain[34]. However, the glass transition is not only determined by the chemical structure of the main chain. Increasing mobility is also caused by increasing spacer length. Additionally, long alkyl chains appropriately sited at the mesogenic moiety may act as lubricants and reduce T_g (see Table 6.2 and Figure 6.11).

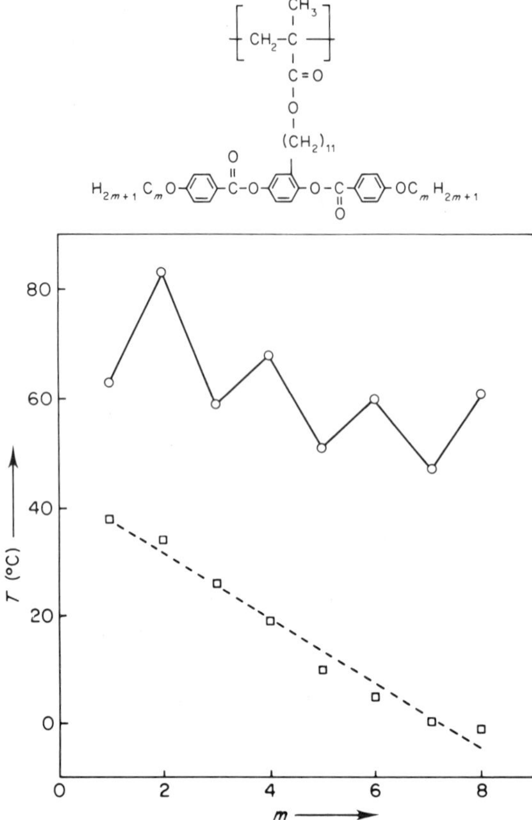

Figure 6.11 Phase transformation temperatures for nematic LC side chain polymers. O = I_c

So far we have considered only linear macromolecules. If the polymerization process is carried out in the presence of a crosslinking agent, three dimensionally crosslinked polymer networks can be synthesized. Above T_g, these polymer networks ('rubbers') also exhibit LC phases up to T_c. Due to the crosslinking, they exhibit viscoelastic properties and can be deformed by applying mechanical stress[34]. The most interesting property of these LC elastomers is the macroscopic orientation of the mesogenic side chains brought about by mechanical deformation.

6.3.3 Properties and aspects relating to applications

LC side chain polymers may exhibit N, Ch and S phases by analogy with LLCs. Because the molecular organization of the mesogenic side chains is similar to that of conventional LLCs, the polymers exhibit the same optical properties. In contrast to LC main chain polymers, where the main chains are anisotropically ordered, LC side chain polymers do not possess any particular mechanical properties. Owing to the linkage of the LC moieties to the polymer back-bone, rotational and translational motions are of course restricted, and this is directly reflected in relaxation processes. In this section will be mentioned some particular properties that offer possibilities for applications.

(a) State of order of nematic polymers. The orientational long-range order of the molecules in the N state is described by the order parameter

$$S = \frac{1}{2}\langle 3\cos^2\theta - 1\rangle$$

The angle θ denotes the mean deviation of *a* molecular axis with respect to the symmetry axis of the orientational distribution function of the molecular axes. This description can also be applied to side chain polymers, if we neglect the polymer main chain and consider only the mesogenic side groups. This order parameter directly reflects the anisotropy of the polymers and has to be considered in the discussion of any anisotropic physical property of a system. Theoretically as well as experimentally, the order parameter has been intensively investigated for LLCs. The question arises whether the order parameters of polymers differ from those of the corresponding monomers. In Figure 6.12, a typical example is shown where the temperature dependence of S for a monomer system is compared with that of corresponding polymers having different spacer lengths[34]. It should be noted that the polymers exhibit lower values of S at a

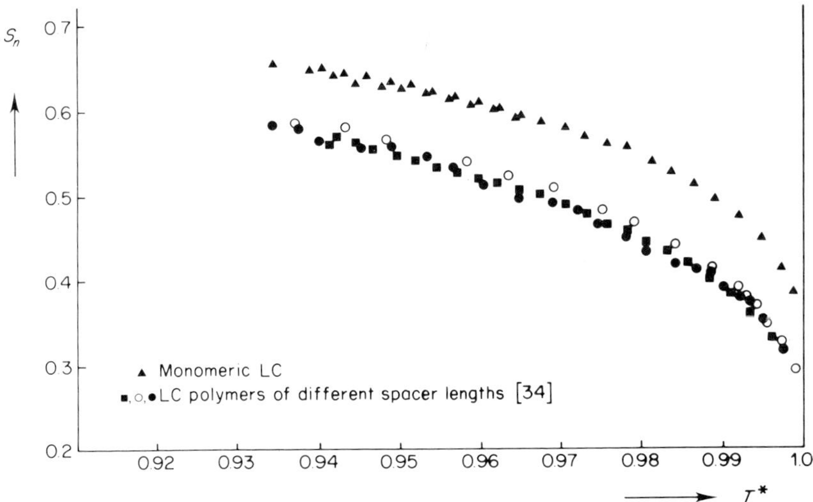

Figure 6.12 Nematic order parameter as a function of reduced temperature T^* ($T^* = T_m/T_c$, where T_c = clearing temperature, T_m = measuring temperature).

fixed reduced temperature, T^* ($T^* = T_m/T_c$, where T_m = measuring temperature, T_c = clearing temperature), while the slope of the temperature dependence curve is very similar. If the order parameter is measured at lower temperatures, S does remain constant at temperatures below T_g. This is one of the most important properties of LC side chain polymers. The LC order freezes in at T_g and remains unchanged in the glassy state of the system. Anisotropic glasses having anisotropic physical properties, e.g. optical properties, can be realized. Applications in optics are therefore obvious for LC side chain polymers[58].

(b) Electric and magnetic field effects. The most important applications of LLC are for display and imaging technology, as outlined in Chapters 3 and 4. An impressive volume of work has been done to optimize the physical properties of LLCs for a range of electrooptical effects. The elastic constants and viscosities of LLCs have a most important influence on the performance of the displays. Without necessarily doing any experiments, it can be stated that LC side chain polymers do not meet the requirements for displays used today, because of their high viscosity. Depending on the degree of polymerization, their viscosity is 4–5 orders larger than that of LLCs[59]. This gives rise to very long response times.

Nevertheless, LC side chain polymers do display the same electrooptical effects and magnetic field-induced orientations as LLCs. Assuming a flexible spacer of sufficient length ($> 4\ CH_2$— groups), the elastic constants of polymers are of the same order[34] as for LLCs. The critical field strengths for the reorientation processes are therefore directly related to the anisotropy of the dielectric constants or the anisotropy of the diamagnetic susceptibilities of the mesogenic monomer unit of the polymer. The anisotropy of the dielectric constants can be varied by changing the chemical constitution of the mesogenic side chains in the manner described in Chapter 3.

The combination of electric or magnetic field-induced orientation in the LC state and the durable storage of the alignment in the glassy state of the polymers offers new aspects for application. Optical elements or storage devices are certainly conceivable[35, 60].

6.4 Conclusions

LC polymers offer a new class of materials, which combine the anisotropic physical properties of the LC state with the characteristic properties of polymers. Depending on the molecular constitution, specific properties result: LC main chain polymers exhibit outstanding mechanical properties, while LC side chain polymers offer field effects and optical properties similar to LLCs, and the realization of anisotropic glasses for optical applications.

Knowledge of LC polymers is still in its infancy compared with that for conventional liquid crystals and conventional polymers. Forthcoming theoretical and experimental studies will, however, undoubtedly extend our perspectives of these interesting materials.

6.5 References

1. D. Demus, H. Demus and H. Zaschke, *Flüssige Kristalle in Tabellen*, VEB Deutscher Verlag für Grundstoffindustrie, Leipzig, German Democratic Republic (1976); D. Demus and H. Zaschke, *Flüssige Kristalle in Tabellen*, Vol. II, VEB Deutscher Verlag für Grundstoffindustrie, Leipzig, German Democratic Republic (1984).
2. L. Onsager, *Proc. New York Acad. Sci.*, **51a**, 627 (1949); A. Ishihara, *J. Chem. Phys.*, **19**, 1142 (1951).
3. P.J. Flory, *Proc. R. Soc. London, A*, **234**, 60 (1956).
4. C. Robinson, *Trans. Faraday Soc.*, **52**, 571 (1956); *Tetrahedron*, **13**, 219 (1961).
5. S.L. Kwolek, US Pat. 3,671,542 (1972).
6. A. Roviello and A. Sirigu, *J. Polym. Sci., Polym. Lett.*, **13**, 455 (1975).
7. W.J. Jackson and H.F. Kuhfuss, *J. Polym. Sci., Polym. Chem. Ed.*, **14**, 2043 (1976).
8. R.W. Lenz, *Organic Chemistry of Synthetic High Polymers*, Wiley–Interscience, New York (1967).
9. J.J. Jin, S. Antoun, C. Ober and R.W. Lenz, *Br. Polym. J.*, **12**, 132 (1980).
10. A. Ciferri, 'Rigid and semirigid chain polymeric mesogens', in *Polymer Liquid Crystals*, ed. by A. Ciferri, W.R. Krigbaum and R.B. Meyer, Academic Press, New York (1980).
11. S.G. Cottis, J. Economy and L.C. Wohrer, Ger. Pats. 2,248,127 (1973) and 2,507,066 (1976).
12. J.J. Kleinschuster, T.C. Pletcher and J.R. Schaefgen, Ger. applications 2,520,819 and 2,520,820 (1975).
13. T.C. Pletcher, US Pats. 3,991,013 and 3,991,014 (1976).
14. S.P. Elliot, Ger. Pat. application 2,751,585 (1978).
15. C.R. Payet, Ger. Pat. application 2,751,653 (1978).
16. R.S. Irwin, US Pat. 4,188,476 (1980).
17. G.W. Calundann, US Pat. 4,161,470 (1979).
18. D.G. Gray in *Polymeric Liquid Crystals*, ed. by A. Blumstein, Plenum Press, New York (1985), p. 369, and literature cited therein.
19. Ch.K. Ober, J.I. Jin and R.W. Lenz, *Adv. Polym. Sci.*, **59**, 103 (1984).
20. L.L. Chapoy (ed.), *Recent Advances in Liquid Crystalline Polymers*, Elsevier Applied Science Publishers, London and New York (1985), and literature cited therein.
21. A. Blumstein, S. Vilasagar, S. Ponrathnam, S.B. Clough and R.B. Blumstein, *J. Polym. Sci., Polym. Phys. Ed.*, **20**, 877 (1982).
22. W.J. Jackson and H.F. Kuhfuss, *J. Polym. Sci.*, **14**, 2043 (1976).
23. A. Blumstein, *Polym. J.*, **17**, 277 (1985).
24. P.J. Flory, *Adv. Polym. Sci.*, **59**, 1 (1984).
25. R.W. Lenz, *Faraday Discuss. Chem. Soc.*, **79**, paper 2 (1985).
26. R.W. Lenz, *Polym. J.*, **17**, 105 (1985).
27. E. Chiellini and G. Galli in *Recent Advances in Liquid Crystalline Polymers*, ed. by L.L. Chapoy, Elsevier Applied Science Publishers, London and New York, (1985), p. 15, and literature cited therein.
28. C.K. Ober, J.-I. Jin and R.W. Lenz, *Polym. J.*, **14**, 9 (1982).
29. W. Weissflog and D. Demus, *Crystal Res. Technol.*, **19**, 55 (1984).
30. K.F. Wissbrun, *Br. Polym. J.*, **12**, 163 (1980).
31. M.G. Dobb and J.E. McIntyre, *Adv. Polym. Sci.*, **60/61**, 61 (1984).
32. A. Blumstein, R.B. Blumstein, M.M. Gauthier, O. Thomas and J. Asvar, *Mol. Cryst. Liq. Cryst. Lett.*, **92**, 87 (1983).
33. G. Ronca and D.Y. Yoon, J. Chem. Phys., **83**, 373 (1985), and literature cited therein.
34. H. Finkelmann and G. Rehage, *Adv. Polym. Sci.*, **60/61**, 99 (1984), and literature cited therein.
35. V.P. Shibaev and N.A. Platé, *Adv. Polym. Sci.*, **60/61**, 173 (1984), and literature cited therein.
36. C. Casagrande *et al.*, *Mol. Cryst. Liq. Cryst.*, **113**, 193 (1984).
37. L. Strzelecki and L. Liebert, *Bull. Soc. Chim. Fr.*, 597 (1973).
38. A. Blumstein and E.C. Hsu, *Liquid Crystal Order in Polymers*, ed. by A. Blumstein, Academic Press, New York (1978), p. 150.
39. V.P. Shibaev and N.A. Platé, *Polym. Sci. USSR (Engl. Transl.)*, **19**, 1065 (1978).
40. M. Portugall, PhD thesis, Mainz, West Germany (1981).
41. B. Hahn, J.H. Wendorff, M. Portugall and H. Ringsdorf, *Colloid Polym. Sci.*, **259**, 875 (1981).
42. F. Cser, K. Nyitrai, J. Horvath and Gy. Hardy, *Eur. Polym. J.*, **21**, 259 (1985).

43. B. Reck and H. Ringsdorf, *Makromol. Chem., Rapid Commun.*, **6**, 291 (1985).
44. H. Finkelmann and G. Rehage, *Makromol. Chem., Rapid Commun.*, **1**, 31 (1980).
45. H. Ringsdorf and A. Schneller, *Makromol. Chem., Rapid Commun.*, **3**, 557 (1982).
46. P.A. Gemmell, G.W. Gray and D. Lacey, *Mol. Cryst. Liq. Cryst.*, **122**, 205 (1985).
47. C.M. Paleos, S.E. Filippakis and G. Margomenou-Leomidopoulov, *J. Polym. Sci., Polym. Chem. Ed.*, **19**, 1427 (1981).
48. P. Keller, *Makromol. Chem., Rapid Commun.*, **6**, 707 (1985).
49. H. Stevens, G. Rehage and H. Finkelmann, *Macromolecules*, **17**, 851 (1984).
50. E. Perplies, H. Ringsdorf and J.H. Wendorff, *Ber. Bunsenges. Phys. Chem.*, **9**, 921 (1974).
51. H. Benthack-Thoms and H. Finkelmann, *Makromol. Chem.*, **186**, 1855 (1985).
52. H. Finkelmann, H.J. Kock and G. Rehage, *Makromol. Chem., Rapid Commun.*, **2**, 317 (1981).
53. H. Ringsdorf and A. Schneller, *Br. Polym. J.*, **13**, 43 (1981).
54. H. Finkelmann, J. Koldehoff and H. Ringsdorf, *Angew. Chem., Int. Ed. Engl.*, **17**, 935 (1978).
55. V.P. Shibaev, H. Finkelmann, A.V. Kharitonov, M. Portugall, N.A. Platé and H. Ringsdorf, *Vysokomol. Soedin, Ser. A*, **23**, 919 (1982).
56. N.A. Platé, R.V. Talroze and V.P. Shibaev, *Pure Appl. Chem.*, **56**, 403 (1984).
57. F. Hesel and H. Finkelmann, *Makromol. Chem.* (1987), in press.
58. H. Finkelmann and H.J. Kock, *Disp. Technol.*, **1**, 81 (1985).
59. V.P. Shibaev, V.G. Kulichikhin, S.G. Kostromin, N.V. Vasileva, L.P. Bravermann and N.A. Platé, *Dokl. Akad. Nauk SSSR*, **263**, 152 (1982).
60. H.J. Coles and R. Simon, *Recent Advances in Liquid Crystalline Polymers*, ed. by L.L. Chapoy, Elsevier Applied Science Publishers, New York, (1985), p. 323.

Index